8/00-20 (12/99)

Collection Management

| 5-01 | 21 | 2-01 |
|------|------|------|
| 2/04 | 30 | 8/03 |
| 1/06 | 30-1 | 7/03 |
| 8/07 | 32-1 | 1/07 |
| 9/09 | 34-1 | 3/08 |
| 5 12 | 37-1 | 1/11 |
| | | |
| | | JUL 1995 |

# Robot Evolution
## The Development of Anthrobotics

# Robot Evolution

## The Development of Anthrobotics

*Mark E. Rosheim*

A WILEY-INTERSCIENCE PUBLICATION

**JOHN WILEY & SONS, INC.**

New York / Chichester / Brisbane / Toronto / Singapore

This text is printed on acid-free paper.

Copyright © 1994 by John Wiley & Sons, Inc.

All rights reserved. Published simultaneously in Canada.

Reproduction or translation of any part of this work beyond
that permitted by Section 107 or 108 of the 1976 United
States copyright Act without the permission of the copyright
owner is unlawful. Requests for permission or further
information should be addressed to the Permissions Department,
John Wiley & Sons, Inc., 605 Third Avenue, New York, NY
10158-0012.

*Library of Congress Cataloging in Publication Data:*
Rosheim, Mark E.
    Robot evolution : the development of anthrobotics / Mark E.
Rosheim
      p.  cm.
    Includes index.
    ISBN 0-471-02622-0 (alk. paper)
    1. Robotics.  I. Title.
TJ211.R67  1994
629.8'92—dc20                               94-13687
                                                CIP

Printed in the United States of America

10 9 8 7 6 5 4 3 2 1

To Dr. Richard E. Byrd, friend and mentor

# *Foreword*

*Robot Evolution: The Development of Anthrobotics* is a unique book that is a blend of robotic history, technology, ideas and trends.

Mark Rosheim's book traces the evolution of robotic systems, beginning with the ancient Greeks, Leonardo da Vinci, and Nikola Tesla. The book also depicts early developments in the industrial robotic sector, including the patents of George Devol and the work of Joseph Engelberger at Unimation. Starting with Chapter 2, the author surveys the state of the art for key technologies required for developing an anthrobot, such as robotic actuators, manipulator joint designs, wrists, hands, and related arm technologies. Advances in each of these areas make an anthrobot a closer and realizable goal. Additional developments in related but integral technologies such as legs, power sources, computer, and artificial intelligence/ behavior science support this direction. Examples of early developed systems, like General Electric's Handyman, lend credence to this idea. The pieces are here, but the time has not arrived.

This brief summary of current robotic technology developments serves to educate the community by collecting past work such that future researchers can build from an existing knowledge base. Even the veterans of the field will find interesting information that could be useful to them in the future. Mark Rosheim has done extensive research into significant patents and events relating to the concept of anthrobotics. Anthrobotics is the study of robots that have a human form, and are sometimes referred to as androids. This book is written for the experienced researcher, as well as the general public, that would like to gain an understanding of robotic technologies. It is useful as a general reference book for the researcher, student, or professional.

While some of the views of the author are controversial, I respect his broad knowledge and contributions he has made in robotics. This book is easy reading and professionally done. It serves its purpose as a general reference text that is both interesting and informative.

Wendell H. Chun
Martin Marietta Robotics Group
Co-Chairman SPIE Mobile Robotics Conference

# *Acknowledgments*

I wish to thank the many people that assisted in this undertaking. First my editing and production team that made it all possible. Mary Ann Cincotta, whose editorial contributions were of immense value. Dave Starr's editing, proofing and desktop publishing skills are responsible for the superb production you see here. Artwork was provided by Larry Morstad, Tom Merchant, Lynn Swenson, Todd Lewis, Rick Lund, and Gene Roban. Additional figure credits are listed on page 416.

Sources that contributed and reviewed materials include Wendell Chun at Martin Marietta Astronautics, Cliff Hess at NASA Johnson Space Center, Matt Rosheim at Apertus Technologies, Jim Karlin at Robotics Research, Carl Weiman at TRC, Branimir Jovanovic, Tesla Museum and Carlo Pedretti, UCLA.

I also wish to thank Art Erdman at the University of Minnesota, who by mistaking me for a graduate student in 1978, funded my first robot component and launched my career.

M. E. R.

# *Contents*

## 3    *Wrists*                         *157*

## 4    *Hands*                          *189*

# Robot Evolution
## The Development of Anthrobotics

*I am not afraid of new inventions or improvements, nor bigoted in the practice of our forefathers . . . Where a new invention is supported by well-known principles, and promises to be useful, it ought to be tried.*

*Letter from* Thomas Jefferson *to* Robert Fulton, 1810.

# 1

# *Robots Past*

*Once out of nature, I shall never take*
*My bodily form from any natural thing,*
*But such a form as Grecian goldsmiths make*
*Of hammered gold and gold enamelling,*
*To keep a drowsy emperor awake;*
*Or set upon a golden bough to sing*
*To lords and ladies of Byzantium*
*Of what is past or passing, or to come.*

W.B. Yeats, *Sailing to Byzantium*

From the dawn of time man has been interested in recreating himself by technological means. Paintings and statues by the ancient Egyptians and Greeks comprised the earliest attempts. A logical extension to statues evolved when moving parts for jaws, arms, and legs were added. Puppets and dolls became common entertainment articles. Applications to religious icons and figures for entertainment spurred further development. The entertainment field still attracts a great deal of effort, as seen in the Disney animatronics. As automation becomes more advanced so does the physical and functional resemblance to its flesh and blood counterpart. This book focuses on only the well documented designs, leaving more mythological reports to other hands. Read *Human Robots in Myth and Science* by John Cohen (1967) for more information on this topic.

The quest for the anthrobot has captivated great minds throughout the ages. Leonardo da Vinci, Descartes, and Tesla all were attracted to making mechanical beings, anthrobots. In each age they made use of the technology of their time, risking ridicule or even imprisonment. Perhaps still the greatest technological problem of our age, this search continues to drive scientists to create anthrobots, a mirror unto man.

## Greeks

The beginning of robots may be traced to the great Greek engineer Ctesibius (c. 270 B.C.). He was product of the Alexandrian School (Alexander the Great 356-323 B.C.), located in the western part of the Nile delta, whose mechanical tradition started with Archimedes (Lloyd 1973). Ctesibius applied a knowledge of pneumatics and hydraulics to produce the first organ and water clocks with moving figures. Ctesibius's disciple, Philo of Byzantium (c. 200 B.C.), wrote *Mechanical Collection* (of which the Pneumatics chapter survives) describing Ctesibius's work. Building on Ctesibius's work, Hero of Alexandria (c. 85 A.D.) wrote *On Automatic Theaters, On Pneumatics, and On Mechanics* (Drachman 1948) and is recognized as the greatest Greek engineer. Hero, famous for the invention of the steam engine, presents the first well documented, workable robots outside of mythology. The Greeks' entertainment robots were dedicated, that is, designed for limited, repetitive functions and did not have to perform any more demanding functions than to be seen and enjoyed. This tradition was spread by Marcus Vitruvius Pollio (c. 25 B.C.), a Roman engineer, through his book *On Architecture*, (37 A.D.). Five of his works on pneumatics survive through Arabic translation. They avoided truly complex tasks, such as performing useful work, because the technology of the day simply could not support it, and abundant, expendable slave labor made automation of any kind unnecessary (Usher 1982). Although the Greeks, as evidenced by their anatomically accurate statuary, possessed deep knowledge of the human form, they lacked the core technology necessary to make a true robot. The first generation of machines, machine tools that could cheaply produce high tolerance components such

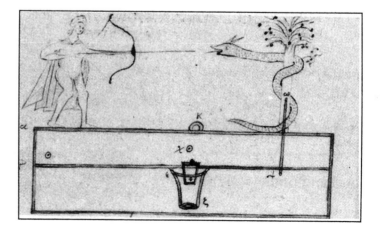

*Figure 1.1  Hercules Shooting Snake*

as envolute gears and bearings, did not exist. The two key ingredients were lacking: need and technology. Success would come only when the limits of technology could match the technical and socioeconomic needs of the times.

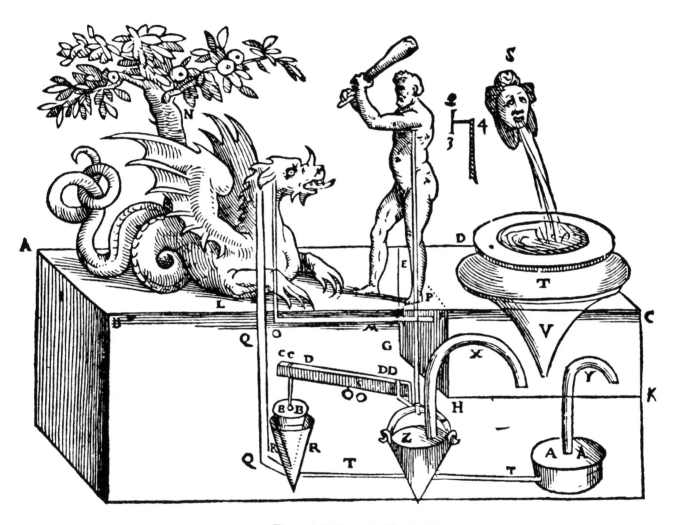

*Figure 1.2 Hercules Slaying Dragon*

Hero's treatises on *Pneumatics and Automatic Theaters* comes to us though the original Greek with the illustrations reconstructed from text. Accordingly, the figures should be viewed with some caution. Thirteenth-century Arab books detail several Greek automata performing various functions, including drinking and pouring. Upon lifting an apple the more action oriented Hercules shoots a snake with a bow and arrow which then hisses (Figure 1.1). The 1598 rendering shows him using a club, the snake replaced by a dragon (Figure 1.2). Water pouring out of head (s) fills basin (T), creating air and water pressure in chamber (c). The water in turn fills vessels (z) and (A) to make the dragon roar and spit; it also actuates the arm of Hercules to strike the dragon's head. A cone valve (R) resets the device.

*Figure 1.3  Birds Made to Sing by Flowing Water*

A modern reconstruction of Hero's singing birds (Figure 1.3) comes from Sigvard Strandh's *The History of the Machine* (1989):

> Water flows the lion's head into a covered container (a), air is then forced out of the container through pipes (b), leading to whistles inside birds. When the water level rises above the top of a pipe (c) covered by a cope - which lets in water at its lower end and is mounted with its top slightly above the top of the pipe - the pipe and cope work together as a siphon, and the water empties from container (a) into another container (d). A similar siphon arrangement (e) is built into this container, which also contains a float (f) to which a rope (g) is tied. The rope is wound once around a rotating pillar (h) and carries a weight (i) at the other end. When the water level - and, thus, the float - rises in container (d), the pillar with its owl rotates until the water level has risen so high that the siphon starts working, and the container is emptied of water. The float then sinks, and the owl rotates back to its original position.

A spectacular 24" (60.96 cm) tall theater (Figure 1.4) was created with such figures, as Hero relates in the treatise *Peri Automatopoitikes* (*Automatic Theaters*) (Geduld 1978):

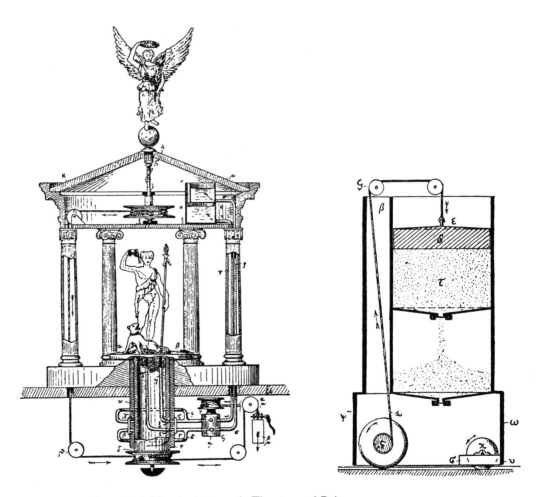

*Figure 1.4 Hero's Automatic Theater and Driver*

A platform, fitted with three wheels, bearing the apotheosis of Bacchus, moved by itself upon a firm, horizontal, and smooth surface up to a certain point and then stopped, at which moment the sacrificial flame burst forth from the altar in front of which Bacchus stood. Milk flowed from this *thysus* [ivy-twined staff], and from the goblet he held streamed wine which sprinkled a panther crouched at his feet. Suddenly, festoons appeared all round the base of the platform, and figures representing the Bacchantes, to the beating of drums and the clanging of cymbals, danced round the temple within which Bacchus was placed. Then the god turned round to another altar, while a figure of *Nike* [personifying Victory], set on the top of the temple, turned in the same direction.... After all these movements had been carried out automatically, the platform returned of its own accord to its starting point.

From the example of Hero we can see that the earliest application of robot mechanics was, like Greek drama, liturgical at root. Another application was as a teaching aid for basic physical laws in pneumatics and hydraulics.

The question arises, why didn't this work continue to develop? The answer lies in the failure of Greek science as a whole to value practical applications. In addition to a tendency to view invention as an art, not a science, there was no socioeconomic mechanism whereby practical scientific work was rewarded. This prevented practical work from continuing, since funding for development was unavailable. A disturbing parallel is seen in today's academics, who are secure in their university positions and unconcerned by practical results, because they are funded by their own institutions and governmental agencies. This artificial system lacks the checks and balances of the real world in which practical applications usually lead to repeated installations and wider dispersal of the ideas embodied in the design. Hero's inventions demonstrate how vulnerable to destruction an idea is when it is confined to a small physical area. Natural disaster or war can eliminate the idea when the one or two prototypes or manuscripts reside in a museum (this was before the introduction of movable type which could produce large numbers of dispersible copies). This is exactly what occurred with the burning of the Alexandrian library and museum, the forerunner of all research institutions, during Julius Caesar's Egyptian campaign in 47 B.C.

**Antikythera Mechanism**

The Antikythera Mechanism (c. 87 B.C.) proves how sophisticated Greek technology was. Possibly constructed by the school of Posidonius (135?-51? B.C.), on the island of Rhodes, it was found by Greek sponge divers in 1900 near the island it is named for (Price 1975). Corroded and encrusted by twenty-one centuries of undersea exposure, X-ray examination revealed a complex mechanical computer in the Archimedean tradition of planetary construction. The differential gear arrangement is composed of thirty-one gears with teeth formed by equilateral triangles. When past or future dates were input through cranks, the mechanism calculated the position of the sun, moon or other astronomical information such as the location of other planets (Field and Wright 1985).

Derek De Solla Price, its main chronicler, argues for its place above mere orrery. "Nevertheless, it does use fixed gear ratios to make these calculations of the soli-lunar calender and it does this more by using pointer readings on a digital dial then by causing a direct geometrical modeling of the planets in space."

The Antikythera Mechanism (Figure 1.5) is the earliest example of a calculating device that processes information through gearing to achieve an answer. Price speculates that it may have been mounted on a statue and located with other automata in a high technology chamber. Pindar (c. 520 B.C.) in his seventh Olympic Ode, speaks of the animated images of Rhodes and Crete,

> The animated figures stand
> Adorning every public street
> And seem to breathe in stone, or
> move their marble feet.

One wonders if the technology of the Antikythera Mechanism was ever directly coupled to the automata of its day in a manner similar to that of the Scribe discussed later in this chapter. Perhaps some day the discovery of other ancient technical wonders will force us to once again reconsider our robotic heritage.

As Price points out, this epoch marked a big change in man's view of himself. Man rivaled the gods as a builder of miniature universes and this conceit is complemented by man's desire to play god by building automata in the human image. The Antikythera Mechanism may have exhibited with automata as a method of displaying both halves of creation. While no one then or now believes that the stars move by clockwork or that ropes move limbs, the Antikythera Mechanism and similar machines fostered the mechanistic philosophy and were important agents for transferring mechanical technology to the present.

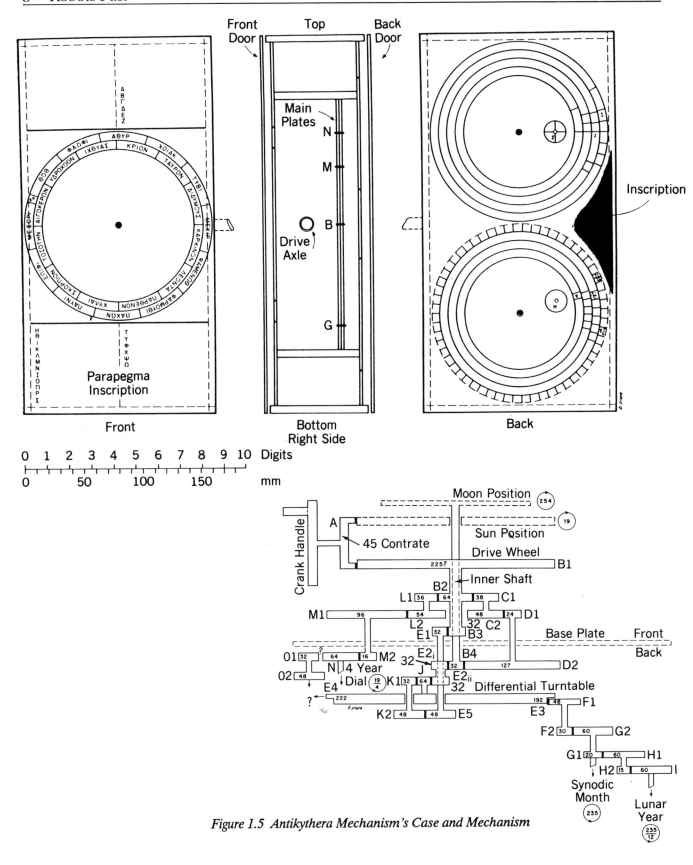

*Figure 1.5 Antikythera Mechanism's Case and Mechanism*

**Arabs**

In the early ninth century the Khalif of Baghdad, Abdullah al-Manum (786-833) commissioned three men, the Banu Musa, to acquire all the Greek texts that had been preserved by monasteries and scholars during the decline and fall of western civilization (Strandh 1989). They produced the large book *Kitab al-Hiyal* (*The Book of Ingenious Devices*) based on the works they collected with embellishments of their own. Over a hundred devices were described. The next significant automation treatise was compiled by Badi'as-Zaman Isma'il bin ar-Razzaz al-Jazari (1150?-1220?). He came from what is today eastern Turkey, northeast Syria, and northern Iraq. Al-Jazari served the Turkman Artuqid dynasty located at Diyar Bakr in southeastern Turkey under Nur ad-Din Muhammad who reigned from 1174-1185 and his successors. They commissioned *The Science of Ingenious Mechanisms*, part compilation or variation of existing designs, and part the author's invention (Hayes 1983). Judging from the detailed descriptions, al-Jazari may have constructed many of the mechanisms. The beautifully illuminated pages are hand painted in multiple colors and adorned with gold leaf. Again hydraulics and mechanics are combined to animate figures, in this case to aid in washing hands. Figure 1.6 shows a female figure by a basin filled with water; when the user pulls the lever, the water drains and the figure refills the basin. This design anticipates the modern flush toilet by several hundred years.

Another intriguing design is the Peacock Fountain (Figure 1.7). Like the previous designs, it is an aid for washing hands of royalty. Pulling a plug on the peacock's tail releases water out of the beak; as the dirty water from the basin fills the hollow base a float rises and actuates a linkage which makes a servant figure appear from behind a door under the peacock and offer soap. When more water is used, a second float at a higher level trips and causes the appearance of a second servant figure—with a towel! Opening the base valve causes both figures to return to their room and the doors automatically close as the water level drops (Geduld 1978).

Unlike the Greek designs, these Arab examples reveal an interest, not only in dramatic illusion, but in manipulating the environment for human comfort. Thus, the greatest contribution the Arabs made, besides preserving, disseminating and building on the work of the Greeks, was the concept of practical application. This was the key element that was missing in Greek robotic science.

*Figure 1.6  Female Figure Pouring Water*

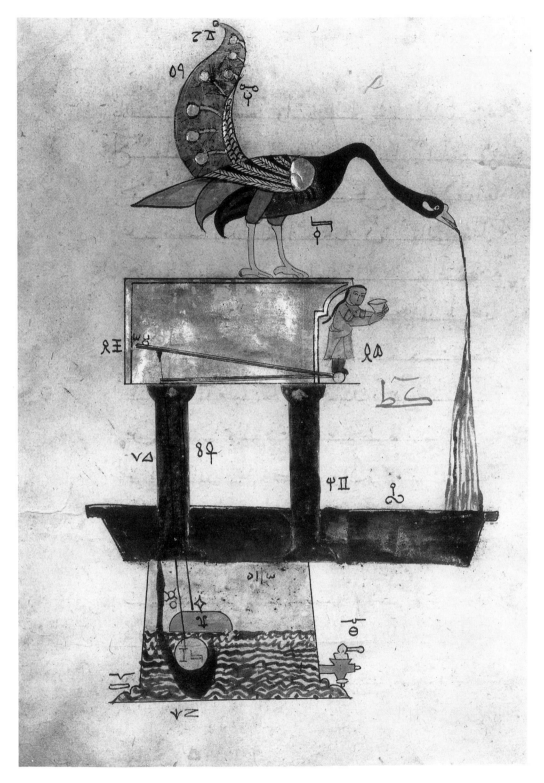

*Figure 1.7 Peacock Fountain*

## Leonardo da Vinci

As mentioned earlier, it appears that the great minds throughout the ages have worked on the problem of robotics, Leonardo da Vinci (1452-1519) among them. The Renaissance revived the interest in Greek art and science but also fostered a desire to compete with the ancients' achievements. Inspired by this, Leonardo followed in Hero's footsteps, studying everything he could obtain on his work. He was actively engaged in verifying Greek reconstructions (Pedretti 1981), an activity that no doubt inspired him to devise water powered organs and clocks equipped with jacquemart," or Jack figures, for striking the hours (Figure 1.8) (Clark 1968). An example of the classic tower clock Jack, circa 1497, is atop the Piazza San Marco in Venice (Malone 1978). Another automaton that Leonardo surely was aware of was the famous Strasbourg cock (Figure 1.9) which was in operation from 1352 to 1789. Part of an elaborate medieval clock, on the hour a cock appeared in the company of twelve figures. Linkages caused it to flap its wings and raise its head, and it crowed three times via a small organ. A strong craftsmanship background acquired in his childhood apprenticeship in drawing, painting, sculpture, and architecture at Andrea del Verrocchio's workshop, as well as self-taught engineering and anatomy knowledge, provided Leonardo with the skills necessary to start where the ancient Greeks had ended. A firm grounding in drafting, anatomy, metal working, tool making, armor design, and sculpture provided him with the ideal skills necessary to build a robot, and funding was provided by the political leaders of his day. He was aware of Hero's works through the translated Arab texts that had become available.

*Figure 1.8 Jacquemart For Striking Bells*

Although no complete treatise has yet been found, new research presents a tantalizing picture. Several pages have been revealed as missing from Leonardo's *Codex Atlanticus* notebook (named for its oceanic size), which some theorize may have contained a spectacular robot study. Moldering in some European library may yet be the lost notebook or pages on robotics or kinesiology (Panofsky 1940). Leonardo, on the other hand, may have decided to keep such a book obscure for the same reasons he

*Figure 1.9 Strasburg Cock*

wrote in mirror handwriting: to elude competitors who might steal his intellectual property. Likewise, any hardware might have been disassembled. Such creations could have also been perceived by religious zealots of the time as blasphemous mockeries of God's handiwork.

Looking at the well documented background on Leonardo, one sees a superb knowledge of anatomy and an understanding of the functions of tendons and muscles — knowledge acquired by clandestine dissections of human cadavers in Rome. Not satisfied with human anatomy alone, he describes in his notebooks a comparative anatomy study to be undertaken: "You should make a discourse concerning the hands of each of the animals, in order to show in what way they vary." An insight comes from *Leonardo on the Human Body* (Schuman 1983). Obviously engaged in building anatomical models, he makes a note to himself: "Make this leg in full relief, and make the cords of tempered copper, and then bend them according to their natural form. After doing this, you will be able to draw them from 4 sides. Place them as they exist in nature, and speak of their uses [Figure 1.10]." Besides models of muscles using copper wires, linen threads or cords, to represent the individual fascicles of muscle, numerous force and cord diagrams have come down to us (Figure 1.11). In his notebooks he also describes making a human heart model out of glass. By 1506-08 Leonardo had made, or at least partially constructed, a wax anatomical model of a whole human figure (Pedretti 1981). After Leonardo's death a visit was paid to his heir by the mathematician and physician Gerolomo Cardano, (inventor of the U-joint) who saw the then intact manuscript collection. In his *De subtilitate*, after remarking that a painter is at once philosopher, architect and dissector, he continues, "for proof there is that remarkable imitation of the whole human body which [I saw] many years ago, by Leonardo of Vinci and of Florence, which was

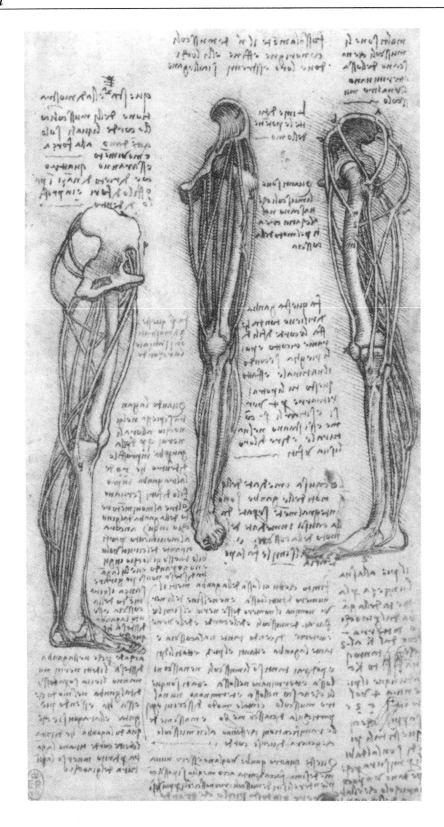

*Figure 1.10  Wire Diagram of the Leg*

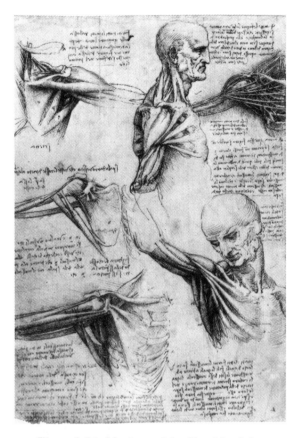

*Figure 1.11 Myology of the Shoulder Joint*

almost complete; [Leonardo was notorious for his incomplete works] but the task required a great master and investigator of nature such as Vesalius" (Schuman 1983).

For his studies on the mechanical possibility of human flight, begun in 1486, Leonardo studied and dissected birds to understand the function of their anatomy. This knowledge, as well as expertise in mechanism design, was put to use in the splendid ornicopter designs showing arm-like linkages articulated by human powered cables. Figure 1.12 shows a multijointed arm operated by pairs of tendons guided by pulleys in an antagonistic relationship. High flexion and distension is achieved by virtue of the several pivots in series. "Mechanism for rotating wing" (Figure 1.13) achieves axial rotation of wing member (h) in a clever yoke design that permits the axial rotation, flexion, and distension of the wing sections. Pulleys at the elbow guide cables to the mechanism on the right. Compare these to the Space Crane in the next chapter. In his full Ornithopter designs the operator's feet pump stirrups which drive cables guided by pulleys to power the wings. The operator's arms are used to assist in driving the wings on the upstroke (Hart 1925, Kemp 1989).

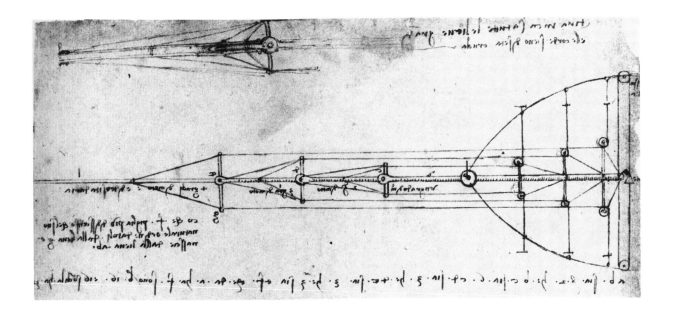

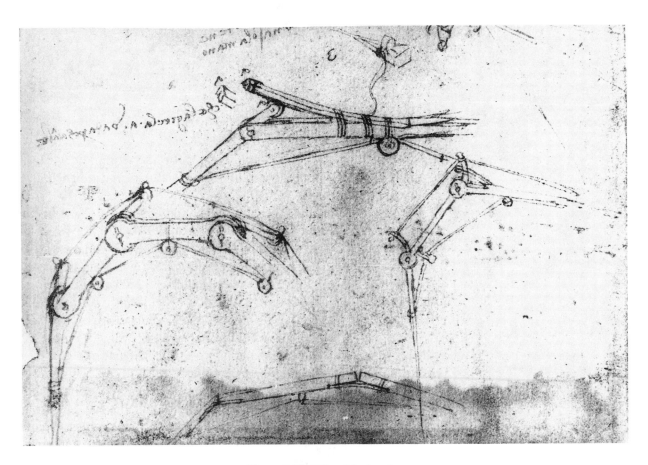

*Figure 1.12 Wing Mechanisms*

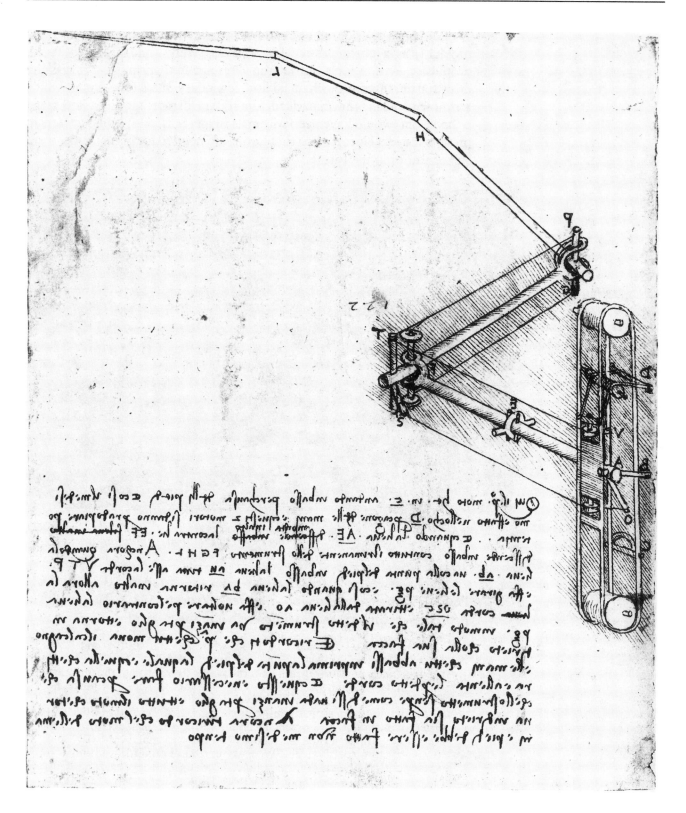

*Figure 1.13  Mechanism for Rotating Wing*

Leonardo considered a version of his leaf spring powered "car" (Figure 1.14) as a power source for actuating wings, and probably as the propulsion system for a automaton. Perhaps leaf spring power was the drive for the famous mechanical lion, legendary in his time, but of which no known drawings have survived. Limited travel (across a courtyard, stage or ballroom) would have been within its range. Sufficient prime movers to provide the ideal locomotion that Leonardo needed for his machines were still centuries in the future.

*Figure 1.14  Leaf Spring Powered Platform for Automata*

Leonardo's robotic fragments have been researched in detail by Carlo Pedretti in his book *Leonardo Architect* (Pedretti 1981). The quality of Pedretti's scholarship makes him worth quoting at length:

> ...We know from Lomazzo in 1584, who said his source had been Francesco Melzi, [one of Leonardo's pupils and heir] that Leonardo used to make "birds, of certain material which flew through the air," that is automatons, and how "once in front of Francis I, King of France, he caused a lion, constructed with marvelous artifice, to walk from its place in a room and then stop, opening its breast which was full of lilies and different flowers." And again, in enumerating the products of Leonardo's technology, Lomazzo records "the way to make birds fly" and "lions walk by force of wheels." We have seen that Leonardo's automaton was a political allegory of the alliance between Florence and the King of France on the occasion of the latter's triumphal entry into Lyons in 1515...

However, nothing more is known of that automaton, and no hint of it is left in Leonardo's manuscripts. But what we do find there are indications of an automaton planned by Leonardo towards the end of the fifteenth century, which was apparently a real robot, of the type of those later constructed by Giannello Torriano for Charles V: a warrior in armour worked by a complex system of cables and pulleys, which articulated his parts, even the jaw, with anatomical accuracy. In f. 366 v-b of the *Codex Atlanticus*, next to the sketch of a mechanical joint for that automaton [Figure 1.15], there is a sketch of the corresponding joint in the bones of a human leg. The few indications remaining are found of the few pages of the *Codex Atlanticus* reproduced here, all datable about 1497. They also include studies of other mechanisms developed in *Madrid MS. I*, which makes us wonder whether the sheets missing from the *Codex Atlanticus* did not contain detailed and spectacular studies of the robot. The existing sketches do not make clear what the motive force consisted of: perhaps springs, as in the early cart design, or counterweights, as in the clocks. Lomazzo's hint on the way of making "Lions walk by force of wheels" could be understood as a reference to a system that Leonardo had already used in the preceding robot project. In a sheet of sketches for the project, *Codex Atlanticus*, f. 216 v-b, we read an incomplete sentence with which Leonardo tried out his pen: "Tell me if ever, tell me if ever anything was built in Rome..." Perhaps this is where the idea of the miraculous architecture of a simulated man comes from, a demonstration of *Anathomi Artifcialis* to the contrasted with Brunelleschi's *Perspectiva Artificialis*, and with which Leonardo, in his subconscious, seems to want to emulate the ancients.

By a curious coincidence, a contemporary of Leonardo's, Benvenuto di Lorenzo della Golpaja, whose writings have preserved a record for us of Leonardo's machines and instruments, wrote the following memorandum at the beginning of one of his codices in Venice: "to remind myself when I go to stay in Rome to make a wooden man who will stand behind a door, and when someone opens the door that wooden man comes to meet him with a stick to hit him on the head, or with a rope to grab him and tie him."

At this stage of technological development we see an effort by Leonardo to equal or excel ancient Greek accomplishments. Still, the core technologies, the ability to machine accurate parts at low cost, and prime movers of sufficient power, are lacking. Although Leonardo did design and probably built machine tools, no single man could bring about in his lifetime an entire industrial revolution. Cheap, expendable labor still abounded, although the working class had achieved a higher status than in Hero's day. The patronage system, which produced spectacular machines like the lion for Francis I, did not provide incentives for producing practical machines. Although movable type printing was available to him, Leonardo did not publish his notebooks. They were copied and circulated however, and did impact the technology of the time: now we start to see early attempts at labor saving automation. The flyer spindle that guides thread winding on spinning wheels is one example of a invention by Leonardo that was adopted; another is locks for boats to pass up and down rivers. He retired to live under the hospitality of King Francis I as "first painter and engineer to the King."

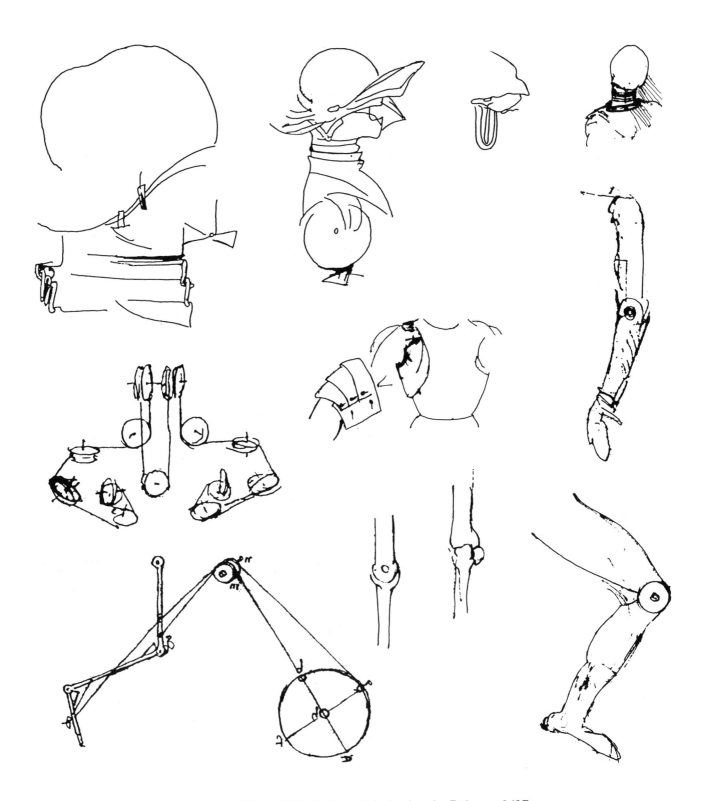

Figure 1.15  Studies of Mechanism for Robot, c. 1497

### The Cittern Player

To provide a feel for one of Leonardo's festival creations I present the beautiful Cittern Player. It was built after Leonardo's death by an unsung Spanish craftsman in the early to mid-1600s (Figure 1.16). It was believed at one time that the robot shown here was the work of the watchmaker Juanelo Torrenano, who was working in Spain. He was also master watchmaker for Kaiser Karl V. Torrenano made several automated human form robots, including a female robot which danced and played the tambourine. Contemporary craftsmen were Tukob Bullman, a watchmaker residing in Kaiser-Hof around 1535, and Christoph Margraf, who built a robot that resided in the art gallery of Kaiser Rudolf II in Prague in 1600. The robots of the "Gondel" in the castle of Ambras at Innsbruck are of the same type and were made in the latter half of the 16th century probably in southern Germany. An 18th century relative of these robots, known as the "Vienna Cittern Player" is exhibited in the Bayrishen National Museum in Munich. Another is in the collection of Charliat in Paris.

The mechanism is made of iron, a wooden hoop casing holds the fall of her skirt. The body, head, hands and feet are made of painted wood. The feet were driven in imitation of walking by a pair of wheels notched for traction. Driven by clockwork, details are unavailable, but I assume that it was spring powered with fusee speed regulation. The figure is 17.32" (44 cm) high and is clothed in linen and silk embroidery. The costume is considered French but could just as well have existed in Spain, Holland or Northern Italy and helps little in determining the origin of the robot. Through the mechanism in the body, the robot moves with small steps and turns its head, while strumming the cittern with her right hand (Kunsthistorisches 1966).

A close cousin from the same period is the Monk (Maurie, 1980). Probably of Spanish origin, it is constructed along the lines of the Cittern Player, a wooden fusse provides speed regulation. Simulated feet protrude from the robe to give the appearance of walking, but wheels provide the locomotion. It is programmed to walk in a 2 foot square, while the head nods, eyes and mouth move, and the right arm beats against his chest while the left arm moves up and down clutching a crucifix. Jesuit missionaries were known to present clocks and automata as gifts to royalty. Perhaps the Japanese Tea Carrying Doll described later in this chapter was inspired by a mechanism similar to the Monk.

Figure 1.16  The Cittern Player

## Jaquet-Droz Mechanical Figures

Well known, the Scribe, Draftsman and Musician display a high degree of anthropomorphism for their time (Figure 1.17). An established clock and watchmaking industry in Switzerland made available the skilled craftsmen and base technology needed to build the robots. Indeed, the Scribe described below features jeweled bearings. Making their public debut in 1774, they were the work of Pierre Jaquet-Droz (1721-1790) and his assistants, chiefly Jean Frederic Leschat. He was born in 1721 at La Chaux de Fonds, Switzerland, to an established family from the city of Neuchatel. Initially he studied theology at Basel University, but later this course was changed to applied mechanics, watch and clockmaking (Bartholomew 1979). One of Pierre Jaquet-Droz's early influences was the duck built by Jacques de Vaucanson and displayed throughout Europe in 1738. Now lost, it was multiple cam driven with one wing containing over 400 parts. Eating, drinking, quacking, splashing its water and even defecating were all functions performed by the duck (Ord-Hume 1973).

Inventor of the wristwatch, the pioneer Jaquet-Droz created lasting examples of the craftsman's art (Jagger 1986). An early innovation was a tiny mechanical singing bird fitted into a snuff box. His mechanical marvels in horology featuring automata prompted an invitation to the Spanish Court (1758-1759). The death of the reigning monarch and the undesirable attention of the Spanish Inquisition, which landed him in the dungeon, cut short this patronage.

Basically entertainment robots, the three figures are programmable through stacked cams. Their movements are in sophisticated and life-like polar coordinates. The Neuchatel Historical Museum booklet, *The Jaquet-Droz Mechanical Puppets* by Alfred Chapuis and Edmond Droz (Chapuis 1956), discusses the first member of the mechanical family as follows:

> The Scribe is a tiny character, a child of about three years old, seated on a Louis XV stool. His right hand holds a goose quill and his left hand rests on a little mahogany table. When he writes his eyes follow the tracing of each letter and his attitude is attentive; his rather jerky movements nevertheless appear absolutely natural.

> The Scribe's mechanism is extremely complex, [Figure 1.18] much more intricate than those of the other two puppets. Pierre Jaquet-Droz must have been faced with very difficult problems: the major one being how to lodge the entire mechanism inside this child-sized body and how to make the elbows and the arms command the movements of the wrists. [Power is supplied by a large spring and regulated by a fly governor.]

> There are two distinct sets of wheel works. An ingenious system sets them alternately in motion until, without any interruption, the last full stop has been accomplished, thereby automatically bringing the whole machinery to a halt.

*Figure 1.17  The Jaquet-Droz Scribe*

The first mechanism is situated in the upper half of the body. It propels a long cylinder on a vertical axis composed of three sets of cams, each of which controls the levers which, in turn, direct the movement of the Scribe's wrist in the three fundamental directions.

In this way the quill not only travels on a single plane but is equally capable of tracing the upstrokes and downstrokes of each letter, as in correct penmanship. Each turn of the cam forms a letter. Then the second mechanism starts: it conveys to the cylinder an upward or downward movement of translation. The length of this stroke is determined by a disc situated in the lower half of the movement; this disc [D] has forty interchangeable steel pegs fixed on its periphery, each peg being set at an angle of nine degrees. Each individual peg impels the cylinder in a determined position corresponding to a specific letter or change of gesture (i.e. beginning a new line, dipping the pen in the ink, etc.).

So it is possible to set the mechanism in such a way that the Scribe will write any desired text of not more than forty letters or signs [in any language].

Many other parts of this intricate piece of machinery deserve description. For instance there is the mechanism which animates the Scribe's head and eyes, and the one which registers the distance when the little table shifts so that there is no irregularity in the spacing of each contiguous letter, whether broad or narrow. Another detail worth mentioning is the apparatus which displaces the dot on the "i" and transforms it into a full stop which automatically brings the mechanical cycle to an end.

*Figure 1.18  The Scribe's Mechanism*

Another version of the Scribe built by Pierre Jaquet-Droz resides in the Peking Museum. A diplomatic gift from England to a Chinese emperor, it features a clock with the figure in European dress. Customized for the eastern market, it included Oriental features and the ability to write in Chinese. In Philadelphia resides an even later version built by Henri Maillardet in collaboration with Henri Jaquet-Droz (Pierre's son and heir) and the craftsman Jean-Frederic Leschot. It was capable of both writing and drawing. After a world tour it was lost for decades, only to be rediscovered in the 20th century and restored. The restorers had assumed it was a girl; imagine how shocked the restorers were when they learned from its messages that it was a boy! This is the first documented case of a robot sex change (Droz & Chapuis 1958).

Other members of this mechanical family included the Draftsman and the Musician. The Draftsman was created between 1772 and 1774, primarily by Pierre's son Henri and Jean Frederic Leschat. Resembling the Scribe, it is capable of rendering portraits, a skill put to the use of flattering royalty. Featuring a simplified, more rational control mechanism derived from the Scribe, the paper in this case is fixed. Its performance is described as "spectacular" with his life-like blowing of dust from the paper, provided by a burst of air from the bellows in his head.

Their sister, the Musician, is a full size adolescent, who actually plays a modified organ in the shape of a clavecin with her articulated fingers. Control is via a studded cylinder similar to the drum from a music box. Linkages pass up through the arms to the articulated fingers. (For more on the Musician's hands turn to Chapter 4.) Realistic breathing is complemented by equally realistic body motion, as she bows to the left and right after she completes her song, which also may have been composed by her inventor (Scriptar-Ricci 1979).

The three figures enjoyed the same royal attention and admiration paid to the wunderkind Mozart, appearing before Marie Antoinette and Louis XVI in Paris. The delighted monarch decorated Henri with an order. When it was announced that it would draw Louis XV's portrait, Henri became so flustered that he caused the Draftsman to write "my doggie" under the king's portrait. One can only imagine the consternation that followed. When later moved by his son Henri Jaquet-Droz to England, the draftsmen drew portraits of George III and Queen Charlotte.

Perhaps the oldest "living" performers of all time, this trio can still be enjoyed at the first of each month in their final home at the Neuchatel Historical Museum. In their old age they have become sensitive to the weather. "The Scribe for instance, cannot tolerate sudden changes of temperature. If the thermometer goes up or down a few degrees, he is in a bad temper. He makes spelling mistakes, forgets to write a letter or neglects his spacing." Fortunately, a fatherly repairman cares for them and makes adjustments to his immortal young/old patients.

The technical problems of building the Scribe's mechanism to perform a high dexterity function and fit inside a small space anticipates the modern problems of robot actuator design and achieving flexible motion in a compact package. A protective enclosure to keep out dust and other environmental hazards is also quite modern. A self-contained power supply was supplied by a spring and regulated by fusee. In addition, the Scribe, like many modern industrial robots, is reprogrammable: it can reproduce whatever text desired within its memory limitation of two to three sentences. However, like all early modern industrial robots programmed by mechanical means, reprogramming is awkward and time consuming, but this is a problem inherent with analog programming methods. The repositioning of the Scribe's hand with respect to the table is an early and ingenious example of using a sensing device to adjust a robot endeffector to irregularities in the work space. Thus the Scribe, though not performing any truly useful work, demonstrates several of the central problems of all robotics systems: power supply, dexterous and compact actuators, protective enclosure of mechanism, stored program memory, as well as sensing and feedback. In its way, the Scribe is the first anthrobot: human-like, flexible, responsive (within limits) to its environment, and capable of performing a human function.

## Japanese Tea Carrying Doll

At approximately the same time that the Scribe and his brother and sister were being manufactured, Japanese craftsmen were developing the first mobile autonomous robot, a tea carrying doll (Figure 1.19). Like Leonardo's lion, the primary effect was for entertainment, in this case, the illusion of a miniature servant (Schodt 1988). Hundreds were built during the Edo, or Tokugawa, period (1600-1867). With a self-contained power supply and a few rudimentary functions designed into it, the "Karakuri" (Japanese for gadget or mechanism) is perhaps grandfather to all autonomous ground vehicles (AGVs). Fourteen inches high, with a mostly wood construction, the Karakuri was held together by wooden pins. The frame was made of cherry, and composite gears of laminated oak and Japanese cedar for strength (Figure 1.20). To regulate the baleen spring, obtained from a whale's mouth, a governor similar to clocks of that era was used. The only metal parts in the device were in this component.

When its spring is wound and it is positioned in the proper direction, it "walks" forward with a cup of tea while bowing its head. After the cup is lifted from its hand, the doll stops and seemingly waits for the cup to be restored. Upon replacement of the cup, the doll performs an abrupt about face and retreats in the opposite direction. A primitive artificial intelligence is displayed.

*Figure 1.19 Japanese Tea Carrying Doll*

Professor Shoji Tatsukawa has recreated the doll from the 1796 three volume manual called *Karakurizui*, or *Sketches of Automata* by Yorinao (Hanzo) Hosokawa. The original text received wide circulation, being printed using wood blocks, which helps to account for its survival. As Tatsukawa has stated:

> It had simple elements of self-control, or sequential control, "in that it could start, stop, and proceed a fixed distance and turn, and return, and it incorporated a governor to run at a fixed speed.... Its use, furthermore, of a cam and spring rather than a bevel gear to solve the difficult problem of changing direction is a truly brilliant idea.

Again we have an example of technology put to an entertainment use for cultural, technological and economic reasons. Like the Greeks, Japanese emphasis was on artistic and intellectual pursuits rather than applying technology to ease man's burden (Hillier 1988). Why bother, when readily controlled human subjects are at your disposal? Technology was viewed as a threat to the well regulated class system of the day—warriors at the top, farmers, craftsmen, and on the bottom merchants. The technologists found a safe outlet for their work, disguising it as art and novelties. Interestingly, modern robots are upsetting the balance between labor and management, but at this date no alarm has been sounded. This is probably because the jobs robots currently perform relieve humans from miserable work conditions for hazardous tasks.

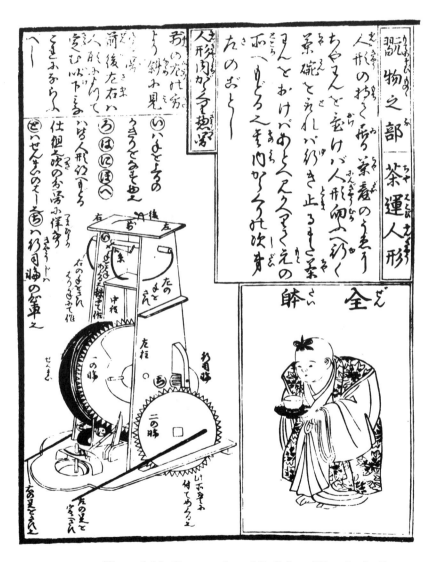

*Figure 1.20  Construction of Doll from "Karakurizui"*

Robots as entertainment devices pioneered by the ancient Greeks continue to be a viable market segment to this day. "Animatronics" developed for the Disney theme parks have entertained millions. Although the technology has improved with the advent of computers and more developed core technologies, the application is the same.

## Nikola Tesla

Nikola Tesla's (1856-1943) growth, from his immigration to America to his becoming one of the industrial revolution's greatest inventors, is the true to life basis for a Horatio Alger story (Tesla 1983). Drawing inspiration from Leonardo da Vinci, Tesla shared his hero's ability at spatial visualization and demonstrated a precocious ability to build model turbines and motors powered by bugs. After a brief attendance at the Austrian Polytechnic School in Graz, Tesla proceeded to educate himself. Arriving in America from Croatia with four cents, a book of his own poems, a couple of technical articles he had written, and some mathematical work, Tesla set out to make his own way. Working for Edison on electrical power improvements, he left under acrimonious circumstances. He eventually obtained backing from George Westinghouse, and developed his alternating current system of generators, motors, and power distribution, which soon replaced Edison's less efficient direct current system (O'Neil 1944).

Tesla's personality and scientific mind, which could solve the most complex problems without pen and paper, are legendary. Tesla led a lonely and celibate lifestyle. Believing that marriage was an aid for the artist or musician, but not for the inventor, he viewed inventing as an occupation that required total dedication without any distractions. In his later career his projects were more of a research and development nature but were billed as commercial ventures. The inevitable crippling cost overruns and lack of immediate returns made financial success elusive. Obviously far ahead of his time yet unable to make his projects commercially successful, he was damned by investors and admired by intellectuals. Had he lived in the day of the government grant, perhaps his life would have ended differently. He died penniless, living on a Yugoslavian government stipend. Tesla has become a kind of patron saint for frustrated, misunderstood inventors. (Cheney 1981)

Tesla's pioneering work in robotics is noteworthy. Inspired by Rene Descartes (1596-1650), in his *Discourse On Method* he constructed a mechanistic philosophy of the soulless machine: "....for the power of self-motion, as likewise that of perceiving and thinking, I held as by no means pertaining to the nature of body...." (Cohen 1967, Torrey 1892). Using himself as a model he theorized at the turn of the century:

. . . I have by every thought and act of mine, demonstrated, and do so daily, to my absolute satisfaction that I am an automaton [the word robot had not been coined yet] endowed with power of movement, which merely responds to external stimuli beating upon my sense organs, and thinks and moves accordingly....

With these experiences it was only natural that, long ago, I conceived the idea of constructing automaton which would mechanically represent me, and which would respond, as I do myself, but, of course, in a much more primitive manner, to external influences. Such an automaton evidently had to have motive power, organs for locomotion, directive organs, and one or more sensitive organs so adapted as to be excited by external stimuli. This machine would, I reasoned, perform its movements in the manner of a living being, for it would have all of the chief mechanical characteristics or elements of the same. There was still the capacity for growth, propagation, and, above all, the mind which would be wanting to make the model complete. But growth was not necessary in this case since a machine could be manufactured full-grown, so to speak. As to capacity for propagation, it could likewise be left out of consideration, for in the mechanical model it merely signified process of manufacture. Whether the automaton be of flesh and bone, or of wood and steel, it mattered little, provided it could perform all the duties required of it like an intelligent being. To do so, it had to have an element corresponding to the mind, which would effect the control of all its movements and operations, and cause it to act, in any unforeseen case that might present itself, with knowledge, reason, judgment, and experience. But this element I could easily embody in it by conveying to it my own intelligence, my own understanding. So this invention was evolved, and so a new art came into existence, for which the name "telautomatics" has been suggested, which means the art of controlling movements and operations of distant automatons.

This principle evidently was applicable to any kind of machine that moves on land or in the water or in the air. In applying it practically for the first time, I selected a boat [Figure 1.21]. A storage battery placed within furnished the motive power. The propeller, driven by a motor, represented the locomotive organs. The rudder, controlled by another motor likewise driven by the battery, took the place of the directive organs. As to the sensitive organ, obviously the first thought was to utilize a device responsive to rays of light, like a selenium cell, to represent the human eye. But upon closer inquiry I found that, owing to experimental and other difficulties, no thoroughly satisfactory control of the automaton could be effected by light, radiant heat, Hertzian radiation, or by rays in general, that is, disturbances which pass in straight lines through space. One of the reasons was that any obstacle coming between the operator and the distant automaton would place it beyond his control. Another reason was that the sensitive device representing the eye would have to be in a definite position with respect to the distant controlling apparatus, and this necessity would impose great limitations in the control. Still another and very important reason was that, in using rays, it should be difficult, if not impossible, to give to the automaton individual feature or characteristics distinguishing it from other machines of this kind. Evidently the automaton should respond only to an individual call, as a person responds to a name. Such consideration led me to conclude that the sensitive device of the machine should correspond to the ear rather than to the eye of a human being, for in this case its actions could be controlled irrespective of intervening obstacles, regardless of its position relative to the distant controlling apparatus, and last, but not least, it would remain deaf and unresponsive, like a faithful servant, to all calls but that of its master. These requirements made it imperative to use, in the control of the automaton, instead of light or other rays, waves or disturbances which propagate in all directions through space, like sound, or which follow a path of least resistance, however curved. I attained the result aimed at by means of a electric circuit placed within the boat, and adjusted, or "tuned" exactly to electrical vibrations of the proper kind transmitted to it from a distant "electrical oscillator." This circuit, in responding, however briefly, to the transmitted vibrations, affected magnets and other contrivances, through the medium of which were controlled the movements of the propeller and rudder, and also the operations of numerous other appliances.

By the simple means described the knowledge, experience, judgment - the mind, so to speak of the distant operator were embodied in that machine, which was thus enabled to move and to perform all its operation with reason and intelligence. It behaves just like a blindfolded person obeying direction received through the ears. The automatons so far constructed had "borrowed minds," so to speak, as each merely formed part of the distant operator who conveyed to it his intelligent orders; but this art is only in the beginning. I purpose to show that, however impossible it may now seem, an automaton may be contrived which will have its "own mind," and by this I mean that it will be able, independent of an operator, left entirely to itself, to perform, in response to external influence affecting its sensitive organs, a great variety of acts and operations as if it had intelligences, I will be able to follow a course laid out or to obey orders given for in advance; it will be capable of distinguishing between what it ought and what it ought not to do, and making experiences or, otherwise stated, of recording impressions which will definitively affect its subsequent actions. In fact, I have already conceived such a plan.

Recollecting the same period some 19 years later in an *Electrical Experimenter* article, he relates:

The idea of constructing an automaton, to bear out my theory, presented itself to me early but I did not begin active work until 1893, when I started my wireless investigations. During the succeeding two or three years a number of automatic mechanisms, to be actuated from a distance, were constructed by me and exhibited to visitors in my laboratory. In 1896, however, I designed a complete machine capable of a multitude of operations, but the consummation of my labors was delayed until 1897...when first shown in the beginning of 1898, it created a sensation such as no other invention of mine has ever produced. In November, 1898, a basic patent on the novel art was granted to me, but only after the Examiner-in-Chief had come to New York and witnessed the performance, for what I claimed seemed unbelievable. I remember that when later I called on an official in Washington, with a view of offering the invention to the Government, he burst out in laughter upon my telling him what I had accomplished... Teleautomata will be ultimately produced, capable of acting as if possessed of their own intelligence, and their advent will create a revolution. As early as 1898 I proposed to representatives of a large manufacturing concern the construction and public exhibition of an automobile carriage which, left to itself, would perform a great variety of operations involving something akin to judgment. But my proposal was deemed chimerical at that time and nothing came of it. (Johnson 1982)

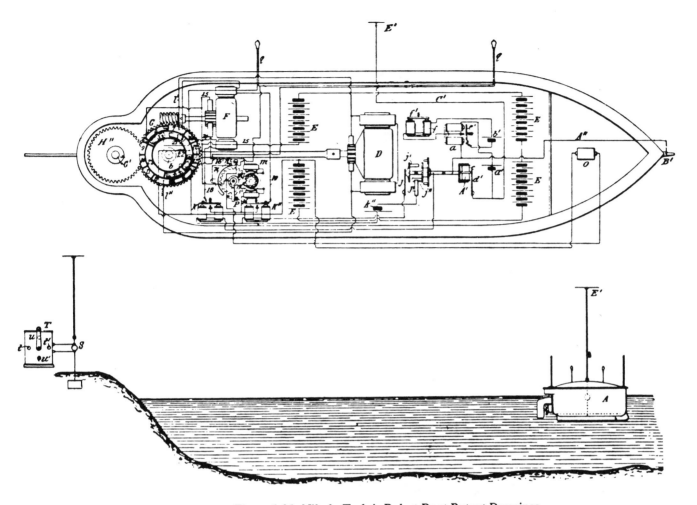

*Figure 1.21 Nikola Tesla's Robot Boat Patent Drawings*

To the amazement of thousands of people gathered at Madison Square Garden, in 1898 Tesla demonstrated a remote controlled, submersible boat (Figure 1.22). It was controlled by coded pulses via Hertzian waves" (i.e. radio controlled). Upon instructions from the audience he would command the boat to turn left, right, submerge, etc. This is the first publicly demonstrated forerunner of all remote controlled machines. Unfortunately, at the time he demonstrated his boat, entertainment of the masses, not continued research funding, was accomplished.

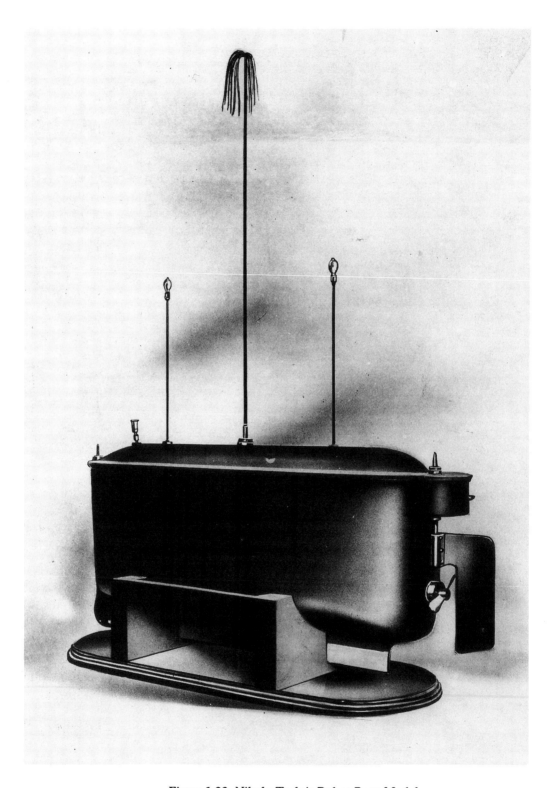

*Figure 1.22  Nikola Tesla's Robot Boat Model*

Tesla's ground breaking work in robotics warrants further study. He wrote later in his life,

> I treated the whole field broadly not limiting myself to mechanics, controlled from a distance but to machines possessed of their own intelligence. Since that time I had advanced greatly in the evolution of the invention and think that the time is not distant when I shall show an automaton which left to itself, will act as though possessed of reason and without any willful control from the outside. Whatever be the practical possibilities of such an achievement it will mark the beginning of a new epoch in mechanics.

The military, the most logical customer for Tesla's boat, was simply not advanced enough in its thinking to appreciate the value of this device. As Andrew Carnegie said "Pioneering does not pay." Years later, during World War II, radio controlled tanks were developed and deployed by the Germans. Operating by radio controlled signals based on musical tones, they were introduced too late to make a decisive impact on the outcome of the war.

In the Tesla Museum in Belgrade, Yugoslavia, the Tesla archives reside, including the original submersible boat model. I have made inquiries regarding other Tesla documents concerning other robot-like designs. What, for example, was the apparatus he demonstrated to laboratory visitors? Was it the boat supported on a frame? Or perhaps a model of an autonomous car? The curator, Branimir Jovanovic, has informed me that no evidence has surfaced on any machine other than the robot boat (Jovanovic, personal communication 1990). As with Leonardo da Vinci, perhaps the future will bring forth some yet unknown work that will further establish Tesla's role as a pioneer of robotics.

In an age when radio controlled toys can be purchased for a few dollars, Tesla's hardware seems perhaps less impressive. However, his sweeping vision of the future is not. In many ways it anticipates Norbert Wiener's Cybernetics which linked biological and mechanical systems in greater detail. (See Chapter 6 for more information.) Tesla was also the first to see a direct analogy between machines and man in their mechanics, senses, and controls. In this he is first to think of robots, not as toys or dedicated devices, but as complex integrated systems. This is an insight, as we will see in Chapters 2 and 6, that has been wasted on many modern researchers.

## RUR

The play *RUR* (*Russums Universal Robots*) was written by Carl Capek and made its 1921 premier in Prague, Czechoslovakia (Capek 1973). Starting as ideal workers, the robots (the play introduced the word robot) have emotions added to them to increase their productivity. A variation on the theme put forth in Mary Shelley's *Frankenstein* recurs when the robots turn on their creators and destroy them. The play ends with an interesting twist: a robot couple is presented as the successors to humans.

The play presents the duality of technology, the threat vs. the promise, through the concept of the perfect worker—the robot. From this point on robots have been perceived as both a savior and a potential enemy of mankind. Since the advent of this play, robots have achieved their own unique identity. This is reflected in the appearance of robots built for fairs, expositions, and movies. No longer made to copy the outward appearance of man, they have become a hybrid—functioning as a human while appearing as a machine.

## Conclusion

In this chapter we have taken a voyage, albeit short, through the history of robots from ancient to modern times. Although both the Greeks and early modern Europeans, such as da Vinci, possessed an unsurpassed knowledge of human anatomy, they lacked the cultural predisposition or the basic technology needed to produce serious robots. The same can be said of the Japanese of the Edo period. The Arabs, on the other hand, displayed an interest in creating human-like machines for practical purposes but lacked, like other preindustrial societies, any real impetus to pursue their robotic science. The 18th century Swiss were the first to develop a recognizable anthrobot, a human-like device composed of precision machinery and capable of both reprogrammable activity and of responding to environmental cues. That such a device as the Scribe should appear at the dawn of the Industrial Revolution comes as no surprise, given the critical mass that had been achieved in mechanism and precision instruments, as well as the advent of modern science itself. In Tesla we see the full flowering of the Industrial Revolution: the machine dominates human thinking to the extent that it becomes, in Tesla's philosophy of automata, the model for human functioning, rather than the reverse. Finally, *RUR* presents the modern nightmare of a functionally perfect worker which evolves into man's nemesis. The age of robotics has been born.

# 2

# *Robot Arms*

*Man is a tool-using animal. Nowhere do you find him without tools; without tools he is nothing, with tools he is all.*

Thomas Carlyle, *Sartor Resartus*   1836

## Introduction

The advent of modern machine tools in the 19th century and the consequent ability to manufacture precision machines, coupled with the development of the computer in the 20th century, provided the technology to build facsimiles of the muscles and the brain. The industrial revolution that made robots possible also created demands for their use. Gone were the days when a highly skilled craftsman working years for wealthy patrons fabricated the most routine parts by hand. The repetitive, monotonous work that factories introduced created a structured environment ideal for the current generation of robots. The rise of organized labor allowed workers to become less tolerant of hazardous duties. Gone was the class of slaves exploited in ancient times.

A relationship can be seen between modern robotics and two of the biggest mechanical technologies, machine tools and automobiles, that significantly impacted the industrial revolution. The two followed similar evolutionary tracks. When machine tools were first developed, basic concepts such as mass and proportions of lathe and milling machine beds received wide, even ridiculous interpretations. The original architectural style of framing was not rigid enough for machine tools (Figure 2.1), and a rigid, simplified box or hollow frame style became standard (Figure 2.2). Developed in the mid-1800s by the machine tool pioneer Joseph Whitworth (1803-1874), it was based on the human body (Roe 1987) and used tubular and box-like bone structures. The same process can be seen in the automobile. Can one now imagine roller chains and sprockets used to drive the rear wheels (Figure 2.3). Concepts in transmissions, engines, suspensions, and steering all received unusual and often impractical interpretations before finding consistent implementation by trial and error, as well as theoretical work. Following this pattern, robotics is going through a similar evolutionary pattern of rationalization of structures and components. Initially, simple polar and cylindrical coordinate arms were manufactured because of the simpler conception and implementation of

*Figure 2.1  Architecture Style of Machine Tool Framing*

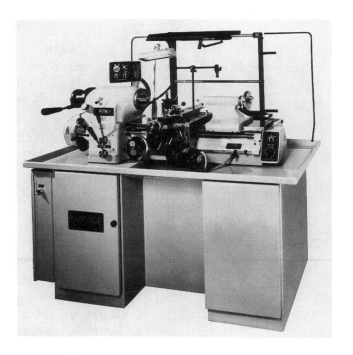

*Figure 2.2  Simplified Box Style of Machine Tool Framing*

*Figure 2.3 Chain Drive for Car*

kinematics known at the time. The level of understanding of kinematics was largely derived from yet simpler machine tools and cranes.

The kinematics of these early robots were also influenced by the linear acting hydraulic cylinder technology. The rise and fall of hydraulic robots is one of the more visible examples of structural evolution brought about by power source changes. Hydraulic power was initially used with commercial robot arms because of the wide availability of general purpose hydraulic actuators, but is being superseded by more efficient electrical systems. Like Edison, the pioneer of direct current power, the pioneers of hydraulics became emotionally wedded to their concepts and are their most dogmatic defenders, even against the less maintenance intensive electric machines. This is ironic, considering their once radical status as progressive innovators.

Evolution of practical mechanisms unique to robotics is a recent development. One highly visible characteristic is the complex three dimensional (3D) shapes of mechanism's components, similar to sophisticated human structures. Cost effective, modular building-block concepts are replacing more difficult to maintain integrated designs. Why should one have to disassemble an entire arm to repair the wrist drivetrain? Machine tools have already proven that modularity has numerous benefits; milling machines are composed of separate modules for each axis. Servicing and repair are simplified because an individual module may be replaced on site and the defective unit shipped to the factory for repair. The availability of more flexible, low speed, high torque, electric drives and standard robotic components, such as wrists and shoulders, is making this possible. Further development of electric linear drives is one area that demands further research. Early designs relied on stock hydraulic drives and components, which produced robots of limited dexterity. Now robots are being shaped by application or tasks, not by their nonrobotic components.

The highly evolved human model continues to be emulated for improved hands, wrists, shoulders, and legs. One of the reasons the human species dominates this planet is the advanced kinematic design that provides highly flexible articulation. Rationalization and standardization of robot design morphologies are progressing to a more anthropomorphic model, as evidenced by the increased use of singularity-free wrists and the large amount of development activity devoted to articulated robot hands. In short, robot structures have progressed from the use of available structures and features to harmonization of form and function. The inspiration provided by the human body gives major impetus to this process of rationalization.

Anatomy and kinesiology provide guidelines for robot designers. Anatomy is the study of animal structure, or morphology. Kinesiology takes the given structure and describes its movements, or kinematics, from the point of view of the physical sciences. Using those insights it is possible to translate human anatomy and kinesiology to robotics. However, replicating animal structure too closely, especially when the replication is of limited practical value, reveals the limits of both the organic design and the inorganic materials. Animal systems are designed for intermediate duty with substantial down time for recuperation. These are highly undesirable characteristics in a factory running two or three shifts. Exact replication of a synovial joint (Figure 2.4) in inorganic materials, substituting an elastomer as the "capsule" for example, would produce break-

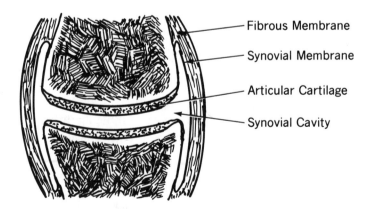

*Figure 2.4  Synovial Joint*

downs and failures. Similarly, inorganic ligaments and tendons that are required to hold these joints together would fatigue, stretch and fail. This is not to deny that someday structurally reinforced elastomers may be created that equal or excel over their animal counterparts. What it does mean is that robot design must consider the nature and limitations of today's medium—inorganic materials—when attempting to parallel animal structures. Some general rules derived from anatomy and kinesiology applicable to robotics are:

- Translate human kinesiology to anthrobotic kinematics and structures

- Remotely locate motors or other drive sources from the joints they actuate

- Provide for self-lubrication

- Enclose all wiring within the structure.

## Robot Arms

This section begins with a discussion of the human shoulder and robotic renditions of it. Design considerations for robot shoulders are examined, with a detailed look at current and future designs. The elbow is then described and its robotic equivalent presented. Types of robotic elbows and applications also are reviewed.

Although human kinematics should be emulated, tissue, tendons, cartilage and bone are not suited for repetitive tasks. Humans apparently were designed for better things. The ever increasing number of legal cases involving carpal tunnel syndrome in meat packing houses, secretarial pools, and other white or blue collar jobs is receiving increased attention. Proliferating lawsuits should be noted by the owners when considering the cost effectiveness of automation (Charlier 1988, Moses 1992).

In general, mechanical robot arms have and continue to follow an evolutionary path similar to that of machine tools; from early, single-task type arms to more versatile, multi-task units. A kind of natural selection process in the business world is suggested by the failure of innumerable robot manufacturers in the 1970s and 1980s. The best technology survives because it has the greatest practical use, pointing the direction for designers to follow if they wish to remain employed. Several radical actuator concepts using inflatable bladders powered by compressed air or hydraulic fluid have been prototyped. Although this type of actuator does find its analog in the male body, unfortunately the reliability and practicality is intrinsically limited by the relativity short life cycle of elastomers. Other, more fantastic visions are painted of direct biological equivalents. However, the organic components of the human arm have proven to be an undesirable design for highly repetitive operations, as witnessed by repetitive motion injuries such as carpal tunnel syndrome and tennis elbow. Robot arms, therefore, must be designed to imitate or exceed the capabilities of the human model while departing, where necessary, from a literal interpretation of the human arm. The use of cables to simulate the action of tendons is one of several examples of a too literal imitation of human anatomy. For example, deletion of a transport mechanism, such as legs, and additional three-dimensional positioning devices, such as the spine, require greater dexterity from a stationary robot arm. To compensate, the robot arm usually is larger and has greater reach than its human counterpart. Designing to the realities of materials and intended application complicates the search for robot arms that are human-like in capability but machine-like in essence.

The chief obstacles to the development of a practical robot design in the ancient and early modern period consisted of an absence of cultural factors and machine tool technology, while the cruxes of robot design in the modern period have been at another level. The advent of machine tool technology made it possible to manufacture precision instruments of all types. The modern frontier has been bounded by an absence of rationalization and standardization of structure/kinematics, power source limitations, as well as control and sensor technology inadequate to direct the enhanced capability of new designs. As these performance barriers are broken, robotics will emerge from its infancy and fulfill humanity's dream of creating the ultimate tool: a machine made in man's own image—an anthrobot.

**The Human Shoulder**

The shoulder is the highest load carrying joint in a human or robot arm. Using nautical terminology, the three degrees-of-freedom corresponding to the human shoulder kinesiology are pitch, yaw, and roll (Figure 2.5). The pitch and yaw axes are perpendicular to the arm and roll motion is in-line with the arm. The shoulder produces the widest range of motion in the human body, 160 degrees horizontal flexion and horizontal extension (yaw), and 240 degrees forward flexion and hyperextension (pitch). Averaging these two numbers gives approximately 200 degrees pitch and yaw. Roll (also called medial and lateral rotation) is 160 degrees, which is increased to 250 degrees by the forearm's additional 90 degrees of roll (Luttgens and Wells 1982). Featuring adaptive load capacity (muscles and bones grow to meet use), the shoulder can lift and carry hundreds of pounds for short durations.

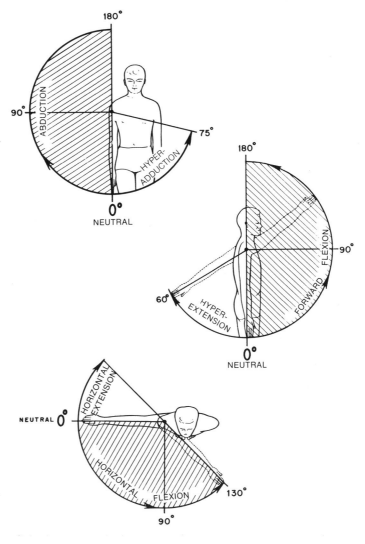

*Figure 2.5  Human Shoulder Range of Motion*

The human shoulder (Figure 2.6) comprises, in part, a ball-and-socket joint providing pitch, yaw, and roll for the humerus (the output). It is attached to the scapula and stabilized by the clavicle. Pitch and yaw motion of the scapula is produced about a separate center point near the rib cage's center. These two joints work in tandem to produce the range of singularity-free motion shown in Figure 2.5.   Figure 2.7 shows the human shoulder modelled progressively as two ball-and-socket joints. The rib cage is generally spherical and the scapula socket shaped. The humerus's ball and cuff is modeled as a smaller ball-and-socket. A kinematic representation is shown in the middle figure and the figure at right shows a further simplification.

Most of the shoulder muscles are located on the torso, not on the shoulder joint itself (Figure 2.8). Such a design provides a high performance joint that is compact, low in mass, with a high load capacity and speed because the joint does not have to carry the mass of the muscles that drive it.   A word should be said about the bone itself. Mainly composed of calcium, it is lightweight, strong, and to a limited extent, flexible: a valuable feature that prevents fracturing from shock loads. The density of the bone

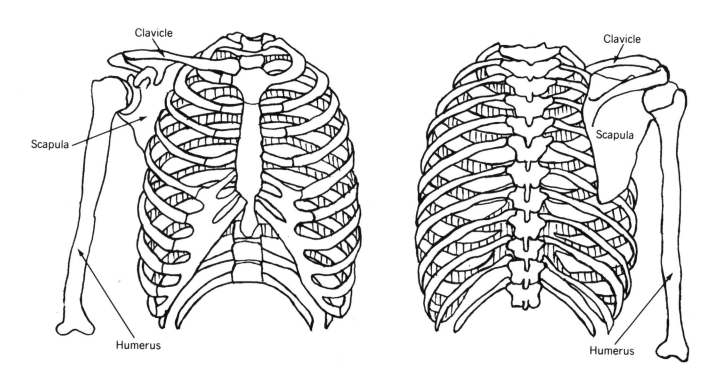

*Figure 2.6 Structure of the Human Shoulder*

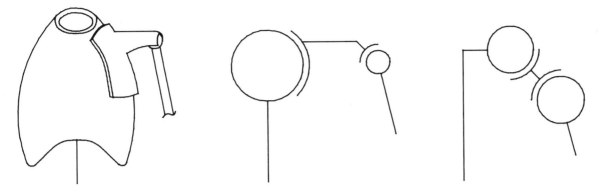

*Figure 2.7 Kinematics of the Human Shoulder*

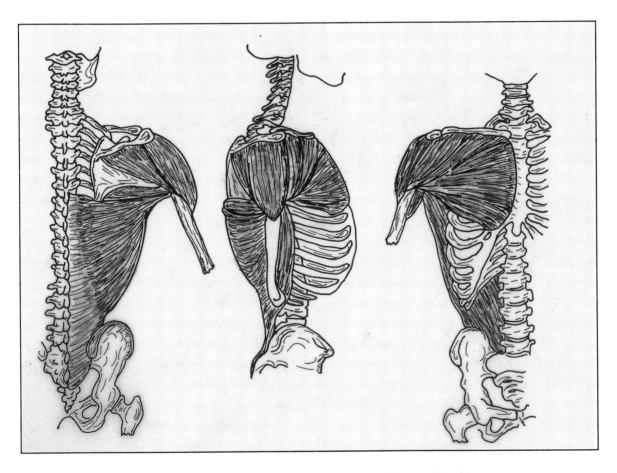

*Figure 2.8 Muscle Location in the Human Shoulder*

varies with the load areas. For robot arms this could be emulated in ceramics, the density perhaps selectively controlled by centrifugal casting processes or chemical milling (see Chapter 6 for more information on ceramics). At the present level of robot materials science, metals remain the substance of choice.

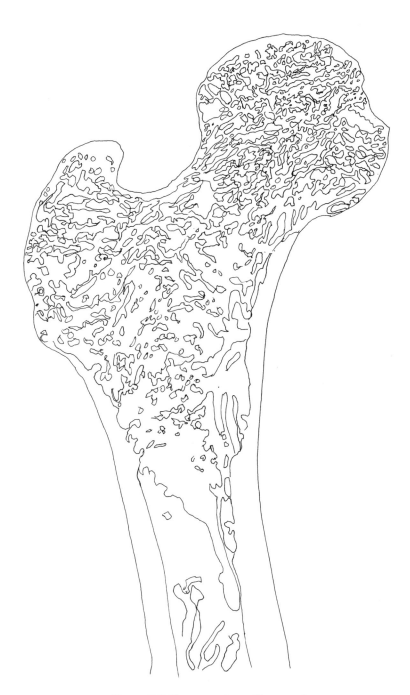

*Figure 2.9 Cross-section of human bone*

## Standards and Measures

Historically, the issue of standardized measurement for robot performance has been considered twofold: first, to measure brute force (the ability to lift a load) and second, to assess precision in factors of repeatability and accuracy. These two concepts reflect historical concepts of robot capability but lack a third element—dexterity. Although at first the main application for the robot arm was the role of the mechanical strongman, the contemporary industrial market is seeking greater dexterity. Dexterity is defined as flexibility of control in the work envelope and range of motion. Flexibility could also be defined as singularity-free motion. Robot designers and manufacturers need to recognize that this shift in emphasis has occurred, and should endeavor to develop standardized measures that give meaningful data about dexterity, not merely range of motion in the various axes, load capacity and precision. A standardized unit of measure is the hallmark of a maturing science and goes hand in hand with the process of rationalization. Without a common conception of relevant robot performance indices, the development of robotics will be hampered by a babel of incompatible terms and design goals. A beginning is the acceptance of the definition of dexterity as the combination of range of motion and flexibility in the work envelope. Dexterity, and measurements in general, must be quantified empirically. For example, pitch-yaw-roll wrists have greater dexterity than roll-pitch-roll wrists. This may be further broken down to individual designs, as for example pitch-yaw-roll wrists with double U-joint structures have more dexterity then single U-joint structures. The addition of dexterity as an integral element of precision will create more precise standards essential for the advancement of anthrobotics. Most industrial robot specifications do not mention singularities in the work envelope and as a result are misleading, if not inaccurate.

Singularity is a major dynamic problem with the shoulders of conventional robot arms caused by kinematics and structural defects. Another word that can be substituted for singularity is jamming, or loss of a degree-of-freedom. Singularity is defined as an area in the joint's work volume which the endeffector must avoid or in which it must abruptly change velocity. The location of singularity in the work volume determines in part the performance of the joint.

Simple pitch/yaw shoulders have been extant since the beginning of teleoperators and industrial robots, as shown in Figure 2.9. Similar to a crane, the base swivels for yaw and the pivot joint above it provides pitch. Because they weren't designed as an integrated system, but as functionally independent units, each joint gives only one type of movement and is functionally independent of the other. This creates singularities or interference with the other joint when they try to operate simultaneously in a multiaxis mode like a human joint. Unlike the singularity-free human shoulder joint, which is basically a double ball-and-socket joint, these crane-like arms have used a simple, additive configuration instead of a multidirectional U-joint such as the ball-and-socket. Fundamental structural and kinematic limitations occur in robotic arms when yaw is achieved by rotating the arm at the base (in essence a roll actuator), with a joint above it providing pitch. A singularity cone results from this degenerative condition (Figure 2.10), caused by the lack of a true perpendicular yaw axis. The singularity cone prevents the endeffector from reaching an important percentage of the work envelope. In a shoulder, singularity occurs when the forearm is vertically aligned with the waist (Figure 2.10a). At this point the forearm is unable to move directly in the planned pitch vector because of the arm's yaw orientation. The arm must first rotate 90 degrees (Figure 2.10b) to get in the pitch mode. A very bulky profile is produced because the pitch motor intrudes in the work envelope. Adding a second roll axis (to decrease the singularity) above the pitch motor introduces the problem of mounting the motor in the upper arm and running wiring through the pitch and roll motor area (Figure 2.11). The second roll motor also introduces an additional load, which further reduces dynamic performance and places an extra strain on pitch and roll motors. Mounting a perpendicular yaw actuator above the pitch joint has been attempted, but also suffers from wire harness management problems. The distended pitch and yaw axes reduce the smoothness of control, and toggling results due to the gap between the axes (see the FTS wrist in Chapter 3 for an example of this). Contrary to popular belief, dexterity is not a function of speed but rather speed is a function of dexterity. How can an arm move at high speed if it is slowed down by singularity zones? A motion may be dextrous and yet very slow, as any ballerina demonstrates.

So little has been done with robot shoulders that this area is a virgin field for designers. For maximum flexibility, a shoulder joint must have a three degrees-of-freedom, pitch-yaw-roll type joint as in the human arm. Decreased load capacity and precision are the trade offs when maximizing flexibility. Placement of arm motors should follow the example of muscle placement in the human arm. Locating motors above the joint improves load capacity and dynamic performance. This places motor weight farther from the end of the arm, where inertial and lever arm effects are the greatest. This approach is only now being applied to designs in which shoulder joint motors are in the trunk, elbow motors in the upper arm, and wrist motors in the lower arm. Table 2.1 provides dexterity ratings for robot shoulders. The table may also be used in a general sense to determine the dexterity of arms.

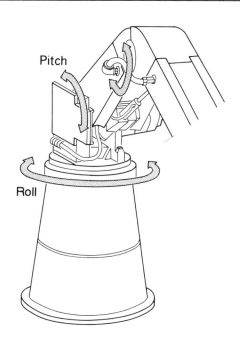

*Figure 2.10  Simple Roll-Pitch Shoulder*

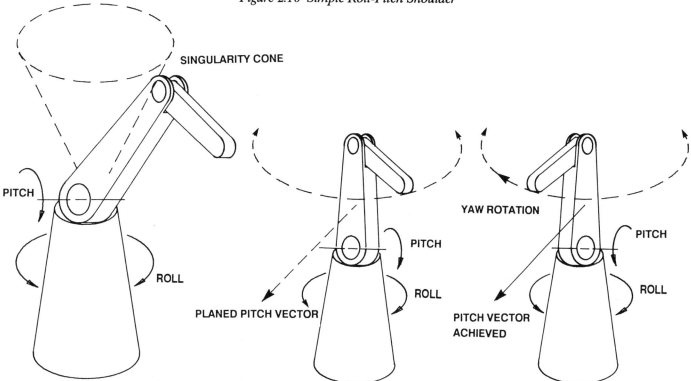

*Figure 2.10a  Singularity Cone*

*Figure 2.10b  Arm Unable to Move in Pitch Vector*

*Figure 2.10c  Roll Rotation Allows Pitch Vector*

Table 2.1  Dexterity Ratings for Shoulders

| Roll-Pitch | Roll-Pitch-Roll | Gimbal, Single Center | U-Joint, Single Center | Double U-Joint |
|---|---|---|---|---|
| 360° Roll 180° Pitch Standard design. Singularity | 180° Roll 209° Yaw 360° Roll Becoming more common. Singularity | 90° Pitch 90° Yaw 360° Roll Same applies to Tripod. Singularity-free | 180° Pitch 90° Yaw 360° Roll Difficult to intergrate roll. Singularity-free | 180° Pitch 180° Yaw 360° Roll Equivalent to human shoulder. Singularity-free |

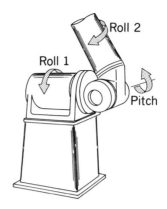

*Figure 2.11 Roll-Pitch-Roll Shoulder*

Another basic need in robotics is a standardized measure of mechanical robot performance. This would assist in rating both existing robots and robots under development. One definition that requires change is output power rating. This should be measured in torque, not in measures of weight attached to the tool plate. Real world applications require the exertion of force at a distance from the joint, rather than the simple lifting of heavy objects whose mass distribution is undefined. Present day robot arm testing places emphasis on the latter and fails to recognize the issue of torque. The result is misleading information about output power ratings. Robot manufacturers, following this misconception, often use lead plates in testing load capacity. One can think of few real world loads like this. Tests should also be developed to measure the ability to manipulate objects of given weights and sizes at varying distances from the power source. In this way it would be possible to accurately assess how much output power was being conserved or lost as a function of drivetrain friction, motor performance, and kinematic form. This "robot dynamometer" could be a valuable development for determining robot efficiency.

Another central problem in robot joint design is that of precision, a term that combines repeatability and accuracy (Figure 2.12). The degree of precision achieved from an endeffector is a function of backlash in the design. Backlash may be defined as an aggregate of sloppiness in all the joints intervening between the endeffector and the power source. Backlash can be visualized as each interface (joint or joint element) introducing discontinuity (slop) between parts. Imprecision (backlash) increases as additional discontinuities are introduced. Backlash in the shoulder is magnified the farther the endeffector is placed from the source of the backlash. Such effects are further magnified and complicated by the addition of other interfaces. Typically the movements of the endeffector are calculated not from the position of the endeffector itself, but from the fixed position of the base joint. The end result is a robotic device that has difficulty orienting itself precisely with respect to the workpiece.

Robot arms are particularly prone to backlash because of the relatively large forces acting on a number of joints and other mechanical interfaces. Care must be taken to design joints that while flexible, are able to control or compensate for backlash. Several techniques for eliminating backlash have been developed (Chun & Brunson 1987, Chun 1986, Yasuoka 1986, Weyer 1987, Nagahama 1990, Baily 1989, Gottschalk 1992).

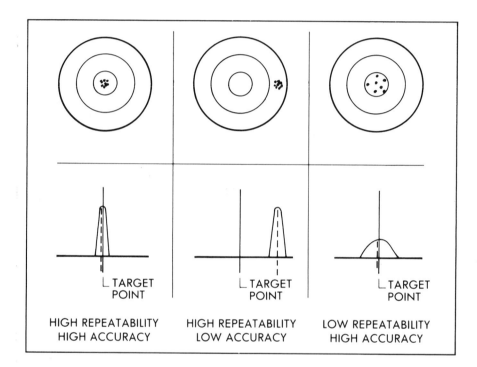

*Firing patterns (top) and distribution curves (bottom) illustrate two important performance specifications: accuracy (a robot's ability to move to the target) and repeatability (a robot's ability to return to the same point).*

*Figure 2.12 Repeatability and Accuracy Measurements*

**The Human Elbow**

The human elbow provides extension, retraction, reach-around, and angular reorientation of the wrist and hand. It also allows the arms to be used as crude grippers, such as when the arms are wrapped around a sack of groceries, posed to carry a load of firewood, or giving a hug (Figure 2.13).

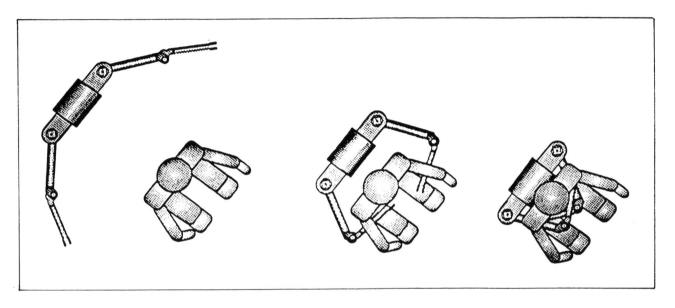

*Figure 2.13 Elbows Used for Grasping*

The human elbow (Figure 2.14) consists of the double-rounded terminus of the humerus, which mates with the ulna and radius to form a pivot joint that is held in place by tendons and muscles. Muscles in the upper arm connect to the forearm and provide power. The synovial fluid that lubricates the joint is contained by the capsule, a membrane enclosure, that encompasses the joint. Normal range of motion for the elbow is 150 degrees, but 10 extra degrees are available by hyperextension (Figure 2.15).

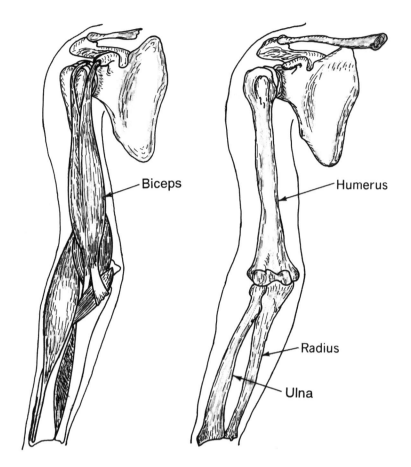

Figure 2.14  Structure of the Human Elbow

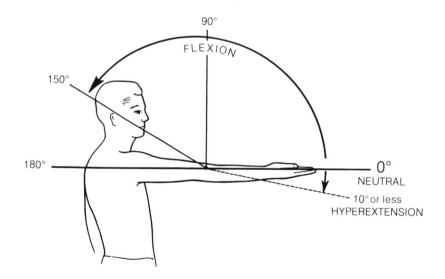

Figure 2.15  Elbow Joint Range of Motion

**Types of Robot Elbows**

Two types of robotic equivalents for elbows have been designed: revolute and telescoping (Figure 2.16). Both designs permit extension and retraction of the endeffector and, with the exception of the telescoping elbow, allow angular repositioning. Revolute elbows may be subcategorized by their type of drivetrain: intermediate, remote or direct drive (Figure 2.17). The intermediate drive elbow has a linear drive (usually a ball screw or hydraulic cylinder) placed between the forearm and upper arm. This is analogous to the human arm's design in that the ball screw or hydraulic cylinder emulates the action of the muscles of the humerus and forearm on a pivot joint. Sharing merits and demerits with the direct and remotely driven elbow, it is a compromise that has been widely used in commercial robots. Other variations include the 4-bar linkage designs by Tesar (Tesar 1989). The remotely powered elbow is driven by linkrods, chains, bevel gears or bands connected to a power source in the upper arm or base of the robot. More complicated than the intermediate, it incurs an increase in backlash and cost. This may be worth the price to users who require a centralized power source in the robot's base for explosion-proofing or in applications that require power source modularity.

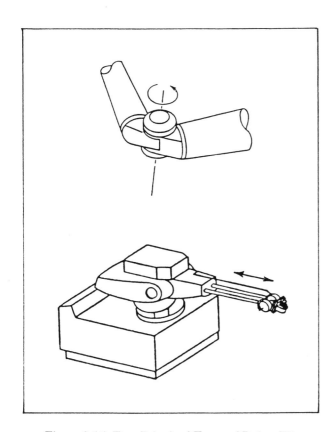

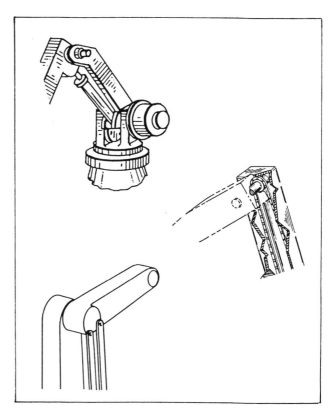

*Figure 2.16 Two Principal Types of Robot Elbows*     *Figure 2.17 Three Types of Revolute Elbow Drivetrains*

The direct drive elbow places the motor directly on the joint, with speed reducers directly coupled to the joint. While lower in cost and similar in design to the remote and intermediate designs, the penalty for this simplicity is an increase in weight and bulkiness at the joint, which decreases speed and load capacity. This direct drive elbow continues to be used by designers who desire low backlash, simplicity of design, and can tolerate its negative qualities.

A major variation from the human elbow is the telescoping elbow. Used since the earliest days of robotics, it provides extension of the forearm without the need for a rotary joint. Simple to construct, its minimal profile provides extension into cavities. The telescoping elbow consists of one or more nested tubes, rods, or extrusions driven by linear actuators. Features include simple construction and consequent low cost with very high stiffness. Heavy loads can be handled by this type of elbow by virtue of its simplicity. Negatives include an inability to reach around objects and generally less flexibility in application.

## Types of Robot Arms

Previous sections have dealt with shoulder and elbow joints as distinct elements. This section addresses shoulders and elbows combined into an arm. A robot arm is a system of structural members that connect a base-mounted shoulder with an elbow joint. The output of the elbow is the output of the arm. Although robot arms are almost universally called "robots," a robotic arm can no more be called a robot than a human arm can be called a human being.

There are five principal types of robotic arms, their designation dictated by their design (Figure 2.18). All consist of simple revolute or prismatic joints configured in the designated categories. The five principal arm types are: the rectangular coordinate (this includes both floor and gantry mounts), polar coordinate, cylindrical coordinate, revolute coordinate, and Self Compliant Automatic Robot Assembly (SCARA). Breaking from this morphology, two new classifications, serpentine and anthropomorphic, have arisen (Figure 2.19) that utilize ball-and-socket or U-joints in their structures. Commercial and experimental examples of all of these configurations will be examined.

Modern robot designers concentrated their earliest efforts on the arm. They chose the arm because its simple movements could be readily reproduced by crane-like forms, and it is the most useful limb for doing work. One of the earliest applications of robot arms included handling hazardous materials. The potentially dangerous effects to humans of toxic chemicals and vapors, heat, radiation, and explosives made this the first and natural choice for robotics. The earliest attempts began in the 1930s with spraypainting, later human powered and human controlled

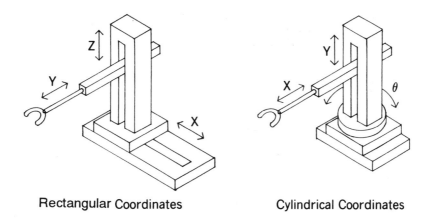

Rectangular Coordinates          Cylindrical Coordinates

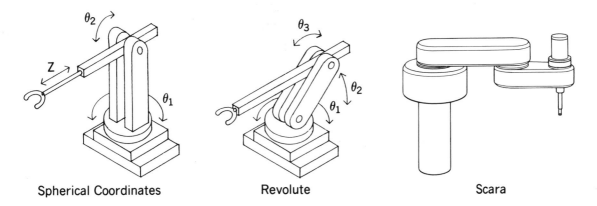

Spherical Coordinates          Revolute          Scara

*Figure 2.18  Five Principal Types of Robot Arms*

teleoperators had rudimentary shoulders, wrists, and hands. Hydraulic robots, directed by early forms of numerical control, were commercialized in the 1960s and started the robotics revolution. By 1973, electric power and electronic control superseded hydraulics in more advanced applications that required superior reliability and increased flexibility in installation and control. Building on this ever broadening base, commercial as well as research organizations have worked toward anthropomorphic robot arms for a diversity of tasks from welding to space station construction.

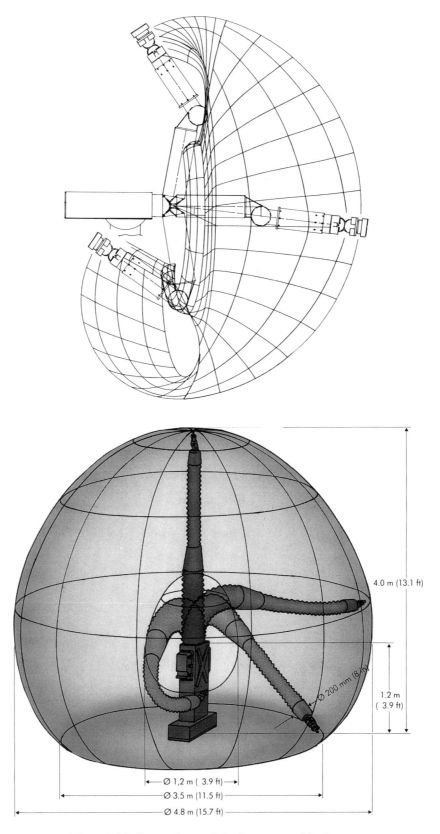

Figure 2.19  Serpentine and Anthropomorphic Arms

## Design Considerations

Early industrial robot structures had inner skeletons sheathed with sheet metal or fiberglass housings. These early structures were kinematically crude as nonanthropomorphic structures were the rule. Current designs utilize a monocoque structure, in which the housing is an integral part of the structure. This exoskeletal approach provides greater stiffness as bending loads on the arm are spread over larger areas. For example, a tube is stiffer or more rigid than a rod of the same weight.

Another of the key concerns in arm design is to provide a drivetrain for the joint. This requirement has been met in a variety of ways. As in the human body, the drive sources for the joint should be placed away from the joint area. When this is done, the joint can be kept compact, while larger and more powerful drives can be employed. Electric drives in the joint itself will never approach remote drives for these reasons. Joint parts should be as few in number and as multifunctional as possible. Multifunctional use of cams, gears, linkages, ball-and-socket joints, and U-joint concepts can accomplish this. For example, a cam might do double duty as a rolling element for a single axis and at the same time decouple itself from the axis. Gear meshes should be kept to a minimum to decrease backlash, and if possible, speed reduction should be achieved with every mesh so there is maximum driveshaft rotation or linkage movement, because this increases the torque and precision conveyed from the driving source.

Hydraulic power sources were originally used to power robot joints because hydraulic technology was readily available, but problems with leakage and maintenance have made an electrically powered and remotely actuated joint the current standard. Transfer of power from the drive sources to the joint may be accomplished by concentric rotating drive shafts, as used in the 3-Roll Wrist, or by push/pull rods similar to those in the Omni-Wrist. Push/pull rods may be driven by linear screw nut actuators in the forearm. Fine leads on the screws give an additional reduction of motor rotation before joint actuation that greatly increases joint control and precision.

Interaction with industry plays a key role in developing designs that meet industry's need. Observations at industrial robot trade shows and contacts with industry leaders through publications and correspondence proves helpful. Unfortunately this type of real world exposure is seldom fostered at American universities. On the contrary, university researchers tend to operate in a world divorced from industry and wonder why their concepts do not interest industry. I have seen many cases where universities have replicated industrial designs and not realized that they were already in use or, worse yet, already abandoned.

## Backdriving

The ability to backdrive an arm, that is to physically move its joints, is useful for force reflection during teleoperation. Backdriving is also valuable for walkthrough programming, such as in spray finishing where the operator physically leads the arm through its task. Backdriving may prevent damage to the joints and drivetrain: if the robot hits an object, the robot "gives." Backdriving requires very low friction; yet almost all electric motors require speed reducers to produce sufficient torque, and the speed reducers produce high friction. Seemingly contradictory, the right combination of low friction and high speed reduction is not easy to find and is seldom achieved with stock components. Two possible approaches are shown in Figure 2.20. Speed reduction may be by screw nut for linear drives or geartrains. In geartrains a minimum number of meshes is necessary to keep backlash and friction to a minimum. Few meshes indicates a single large ratio, such as a chain and sprocket combination where a large reduction is obtained cheaply. See the GRI and GMF robots later in this chapter for an example of this in the shoulder.

Force reflection is the ability of the master operator (whether human or computer) to sense forces remotely. The simplest form occurs in all mechanical teleoperators in which the master is mechanically linked to the slave. Current sensing of motors is used in more elaborate electric arms. In other words, the master operator physically feels what the slave robot does, and the motors of the master and slave act in a synchronous mode. Force sensing through a separate system of load cells frees a designer from the issue of mechanical backdriveability. Refer to the Robotics Research Arm in this chapter and the Sarcos Arm discussion in Chapter 6 for additional information on pioneering work on force sensing.

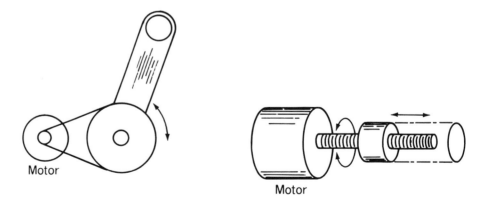

*Figure 2.20 Two Backdriveable Drivetrains*

## Counterbalancing

Counterbalancing reduces the burden on motors by negating the weight of the arm or other limb. Counterbalanced designs may be as simple as wrist motors located on the opposite side of the elbow's fulcrum point to counterbalance the weight of the forearm. Lead weights are used also, but the increased mass adds inertia to the system. More complex designs use springs and air cylinders in various ways to achieve the same effect. The most sophisticated technique is electronically counterbalancing (see Robotics Research's Arm for an example). Advantages to electronic counterbalancing are that an axis may be altered individually, without affecting the other related axes, and backdriving may also be achieved. Mechanical backdriving and counterbalancing have found wide application in spray painting robots because the load does not vary.

The shoulder commonly requires counterbalancing. Figure 2.21 shows a rack and pinion drive preloaded by double springs to counterbalance the weight of the arm (Witterwer 1988). An air cylinder accomplishes the same function in Figure 2.22 (Svensson 1990). More ingenious is the concept of using two elastomeric bushings at the shoulder axis as shown in Figure 2.23 (Young 1989). This is more compact and lightweight. The upper arm is connected to the inner diameter of the bushing and the base to the outer diameter. The torsional windup acts as a spring to counterpoise the arm. Chains are often used to connect the air cylinder or spring to the arm. Although chains are simple to design with, reliability is a problem.

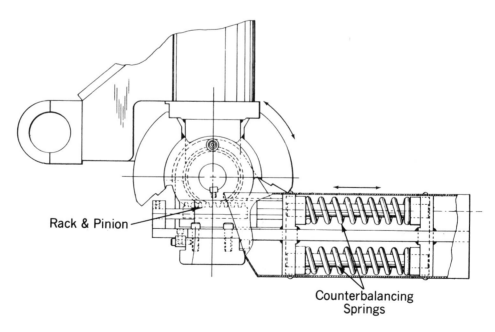

Rack & Pinion

Counterbalancing
Springs

*Figure 2.21  Rack and Pinion Counterbalancing*

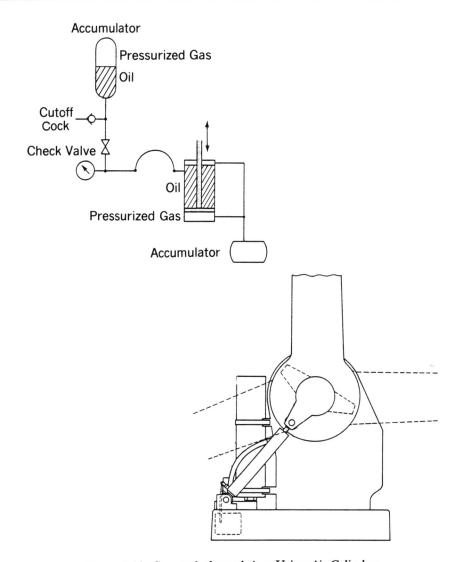

*Figure 2.22 Counterbalanced Arm Using Air Cylinders*

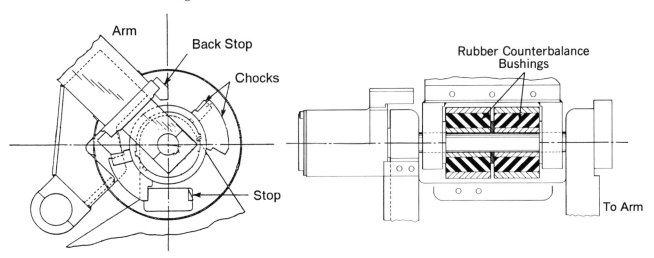

*Figure 2.23 Counterbalanced Arm Using Rubber Bushings*

## Components

A thorough working knowledge of basic components is fundamental to designing robots. For more extensive information see *Robot Wrist Actuators* (Rosheim 1989b) or *Mechanical Design of Robots* (Rivin 1988). Bearings common to all robots are one of the simplest mechanical elements. In my designs I employ a wide variety of types. Conical Timkin type roller bearings commonly used in headstocks of lathes are unsurpassed for high precision and stiffness. They are used in pairs for the center column of the Omni-Wrist described in Chapter 3. Thin section bearings such as the Kaydon type have found application in many robot joints. Their compactness, light weight and reasonable load capacity are valuable features. Because of interference problems and high load capacity requirements, conventional bearings are not desirable in many of my designs. Bearings are engineered and built from scratch, taking advantage of integral raceways machined into housings and other members. This practice invariably raises the eyebrows of engineers accustomed to working only with catalog parts. In addition to higher load capacities, custom bearings with integral raceways are advantageous in situations where parts have relative motion and need to be "keyed" together by the balls. The "keying" increases system reliability by making separation almost impossible. An example of this is found in the outer ring bearings on the Omni-Wrist housing (see Chapter 3).

Other useful types of bearings include linear bearings such as the Thomson Ball Bushing which utilizes recirculating ball bearings in oval tracks. Less well known is the Rotainer manufactured by Lempco Industries. A simple perforated sleeve filled with ball bearings rolls on an inner shaft and outer bushing to provide not only linear but also rotational motion like a bushing.

Ball-and-socket joints have many applications, not only as components, but in modified forms as robots joints themselves. The three basic types are (1) ball with stem in full socket, (2) ball-and-socket with open socket and optional hollow or solid ball, (3) ball-and-socket insert hollow ball.

U-joints, like ball-and-sockets, may be drivetrain elements in robots or modified to form robot joints themselves. The classic form known as the Hooke, Cardan or universal (U) joint is used in the rear wheel drivetrain of cars. Several variations exist including fork-in-groove joints, and the constant velocity U-joints, balls-in-groove (Rezzepa) joints used in front wheel drive cars. Gimbals, which derived from the U-joint, are essentially a hollow U-joint.

After bearings, gears are one of the most basic drivetrain elements in existence. Durable and rugged, they may transmit power at efficiency of up to 98 percent. The simplest and oldest form of gear is the spur gear.

This transfers rotary motion in a single plane and is simple to align and maintain. For greater efficiency and smoother operation the helical gear has angled teeth. Cost is consequently higher because of more elaborate tooth form. Bevel gears transfer motion at angles and must be "set" in two planes. They have been used extensively in the PUMA robots and are the chief maintenance problem. Preload adjustments are difficult and often result in excessive gear wear. The spiral bevel gear, a variation of the bevel gear, can carry a higher load but is predictably more expensive because of the special curved tooth form.

The bane of all gear systems is backlash. Backlash is the play or slop between gears when they mesh. This can be reduced or eliminated through high tolerance gear cutting, or by using split pairs that oppose each other and create a preload.

Speed reducers usually contain gears of a high ratio. The benefit of speed reduction is that torque is increased while the speed is reduced. This is essential in cases involving motors operating at high RPMs. Speed reducers may be as simple as a pair of gears. Harmonic drives are one popular device for high ratio reduction (up to 200:1) and are available in a wide range of sizes and load capacities. Harmonic drives consist of a flexible steel band, a flexspline, with outwardly facing gear teeth that engage a cup with inwardly facing gear teeth. An elliptically shaped wave generator drives the flexspline through bearings to produce a continuous rolling mesh with the cup.

More exotic are the cycloidal type speed reducers made by Dolan-Jenner (Woburn, Mass.) and Sumitomo Machinery Corp. of America. An arrangement of eccentric shafts, rollers, and specially shaped gears produces high speed reduction with very low backlash. High stiffness is also achieved through this type of planetary drive.

Chains and belts are other motion transfer devices, but they, like ball screws, have fallen from favor with robotic customers due to their tendency to stretch and fail unpredictably. Millions of cycles are required in environments where preventive maintenance is not religiously practiced because of multiple shifts or poor management practices.

**Actuators**

Actuators translate power into motion. Electric motors, hydraulic cylinders, pneumatics, and the newer shape memory alloy, as well as polymeric, piezoelectric, and magnetostriction may be considered. John Hollerbach and Ian W. Hunter have written a paper analyzing all of these systems (Hollerbach and Hunter 1991). Tables 2.2 and 2.3 compare the various actuators' performance characteristics. The following is a brief review.

Electric motors have basic features that make them attractive for anthrobots. They may be driven directly by batteries or fuel cells without the need for intermediate pumps, reservoir, or accumulators. A wide variety of sizes are becoming increasingly available. Feedback devices may be mounted directly to the motor shaft. Brush and brushless permanent magnet DC and AC motors are of the greatest interest. Their primary limit is overheating, which causes thermal insulation failure and short-circuiting. Heat is produced as a by-product of dissipating power in the winding to produce torque. Convection aided by fins and water cooling is being researched to dissipate this heat. A high number of RPMs is necessary for maximum efficiency but requires speed reducers to reduce the speed and increase the torque. An added benefit with the high ratio gearbox is the nonbackdriveability that produces inherent self-braking in case of power loss. This is important to prevent damage to both man and machine.

Direct drive motors offer the feature of zero backlash and high mechanical efficiency. However, large diameters are necessary to produce usable torque. Torques great enough for direct driving revolute type arms have not yet been achieved. At present, their application is confined to SCARA and other nongravity loading designs. Brushless motors use electronic commutation, eliminating noisy, less efficient brushes, but their cost is typically four times as great.

Hydraulic cylinders, rotary actuators, and motors were the original actuators for most industrial robots. High torque/mass and power/mass ratios are achieved compared to electric motors. Also, hydraulic actuators are better suited as linear actuators because they are mechanically more simple than electric powered ones. Servo valves meter fluid to the actuators under computer control. The servo loop is closed through feedback devices mounted on the output shafts of the actuators. Double acting hydraulic cylinders are the most popular linear device. Rotary motion up to 270 degrees is provided by vane actuators which are simply an output shaft connected to a fixed vane. The vane is dynamically sealed in a cylinder and forced to rotate by the pressurized differential between the vane and cylinder wall. Hydraulic motors are more complicated and incorporate valving mechanisms to produce continuous rotation.

Table 2.3  Material Properties of Actuator Media

| Actuator | Stress (MPa) | Strain | Strain Rate ($s^{-1}$) | Mechanical Efficiency |
|---|---|---|---|---|
| Electromagnetic | 0.002 | 0.5 | 10 | 0.90 |
| Hydraulic | 20 | 0.5 | 2 | 0.80 |
| Pneumatic | 0.7 | 0.5 | 10 | 0.90 |
| NiTi SMA | 200 | 0.1 | 3 | 0.03 |
| Polymeric | 0.3 | 0.5 | 5 | 0.3 |
| Piezoelectric | 35 | 0.0001 | 2 | 0.5 |
| Magnetostrictive | 10 | 0.0002 | 2 | 0.8 |

Table 2.2  Comparison of Actuator Characteristics

| Actuator | Torque/Mass | Power/Mass |
|---|---|---|
| McGill/MIT EM Motor | 15 N-m/Kg | 200 W/Kg |
| Sarcos Dextrous Arm electrohydraulic rotary actuator | 120 N-m/Kg | 600 W/Kg |
| Utah/MIT Dextrous Hand Electropneumatic servovalve | 20 N-m/Kg | 200 W/Kg |
| NiTi SMA (Hirose et al, 1989) | 1 N-m/Kg | 6 W/Kg |
| PVA-PAA polymeric actuator (Caldwell, 1990) | 17 N-m/Kg | 6 W/Kg |
| Burleigh Instruments inchworm piezoelectric motor | 3 N-m/Kg | 0.1 W/Kg |
| Magnetoelastic (magnetorestrictive) wave motor (Kiesewetter, 1988) | 500 N-m/Kg | 5 W/Kg |
| Human biceps muscle | 20 N-m/Kg | 50 W/Kg |

Negative aspects of hydraulics include the necessary motor/pump, accumulator, filters, reservoir and other support hardware that are seldom shown in photographs of hydraulic systems. Although advancements have been made in manifold mounting to reduce the number of flexible hoses required, routine maintenance still includes replacement of oil, servo valves, and at the minimum the input and output hoses to the system.

Pneumatic actuators, like hydraulics, are advocated for low cost and high speed. The compressible nature of air makes control more difficult, although strides have been made in accommodating this problem. Figure 2.24 shows the Rubbertuator by Bridgestone (Loughlin 1991). Elastomeric tubes are housed in a woven flexible sheath, and when pressure is supplied the "muscle" contracts (Takagi 1986, Sakaguchi 1987). An equally exotic design is employed on the Utah/MIT Hand, which employs a single stage suspension valve connected to a glass cylinder, which contains a graphite piston in a bank of thirty-two. As in the hydraulic system a motor/pump, filters, and reservoir are required to complete the system. Unlike hydraulics, if air leaks the hazard is minimal.

Shape memory alloy (SMA) actuators, as used in the Hitachi Hand described in Chapter 4, are an attempt at a simple high power-to-weight actuator. SMA relies on the phenomenon of an alloy returning to a previous shape after heating. NiTi, known under the trade name Nitinol, is the most popular. Typically it is wound in coils or skeins to increase the actuator's stroke. Normal speed is slower than human muscles; however, experiments using electromagnetic fields have increased the speed, and large forces per wire cross section have been reported.

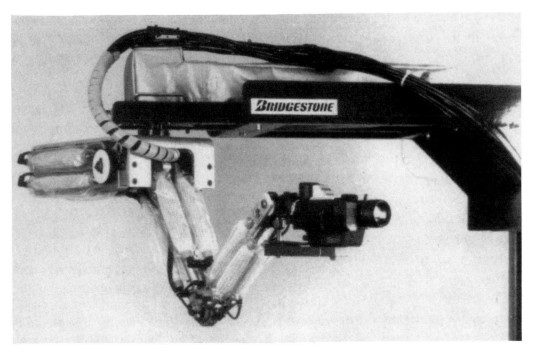

*Figure 2.24  Rubbertuator by Bridgestone*

Polymeric actuators convert chemical or electrochemical energy to mechanical force, and are the closest equivalent to the human muscle. Polyelectrolyte gels, natural and synthetic rubber, and crosslinked collagen are being explored. The Pitman described in this chapter is designed around this technology. One particularly promising polymer is polyvinyl alcohol-polyacrylic acid (PVA-PAA) copolymer. Mechanical efficiency is estimated at 30 percent. Speed and fatigue are problems with this technology, which requires further work to become practical.

Piezoelectric and magnetostriction actuators are based on certain crystals and rare earth alloys that undergo minute expansion when electrical current or magnetic fields are passed through them. The challenge in both these systems is to produce useful mechanical motion out of very short stroke lengths. Linear microstepping or "inch-worm" motion is produced from stick-slip mechanisms. Efficiency is relatively low and practical application has been limited to optical components. Miniature robot arms have been fabricated using this technology in a three degrees-of-freedom wrist (Lee and Arjunan 1987). Magnetostriction inch-worms use brittle Terfenol-D rare earth alloy rods saturated by an electromagnetic coil. Relatively low speeds have been reported with displacement of 50 percent. Hybrid Terfenol-D and piezoelectric motors have been developed in the hope of incorporating the best features of both designs.

Primary factors in the continued development of hydraulic actuators are low cost research and development with impressive performance in speed and load capacity. This I feel obscured maintenance issues and masks conclusions drawn from years of robot actuation experience. Hydraulics leak, whether from failure of hoses or seals or through routine maintenance. Many of the current "new" developments, such as multiple seals and extra passages to capture seal leakage, have been tried in industry in the past but have gained only limited acceptance because of increased cost. Creative, skilled use of electric motor technology, while more difficult to master, is still the most practical means for robot actuation at present. The advantages of clean, maintenance free performance and direct connection to the power source without complicated, inefficient and bulky pumps, reservoirs, filters, and accumulators are but some of the reasons to stay this course.

Electric linear actuators translate a motor's rotary motion into push/pull power. Typically, this is performed by use of a screw nut device with the nut captured in the rotational axis; as the screw turns the nut transverses the screw. A ball screw offers lower friction than a conventional screw nut. Ball screws are losing popularity with robot users and designers because of their tendency to fail; one damaged ball bearing turns the ball nut into a brake. More reliable are planetary type nuts offered by Spirocon and Rolvis. Several threaded rollers in a planetary arrangement engage the screw thread. The housing that encloses the screw provides a race for the rollers to orbit within. An internally threaded load member transfers the load from the rollers to the housing. Another variation of the linear actuator is the Roh'lix and Thomson Ball Nut Linear Drive. Friction is used to create a screw-like effect on smooth shafts using inclined rollers or ball bearings. Ross-Hime Designs has developed a new concept in linear actuators called the Minnac, short for miniature actuator. In place of the conventional recirculating ball screw arrangement, a split self-lubricating screw is used. A high reduction gearbox supplies torque and a highspeed multi-start screw supplies speed. This combination provides an extremely efficient prime mover; over 13 Lbs (5.9 Kg) may be lifted by a 3/4" diameter unit (see Chapter 4 for more information).

## Brakes

Providing fail-safe brakes is an important safety consideration in arm design. Self-braking is not required if the system is counterbalanced or horizontally configured as in SCARA designs. Brakes may be added, as in the NASA FTS design, but they may also add their own reliability problems. A typical electromagnetic brake (Figure 2.25) consists of an electromagnetic coil, armature, brake pad, and rotor. De-energizing the electromagnet (normal-off brake) releases the armature that brings the brake pad into rotor contact. Another option that may be added is a manual release for maintenance purposes. Many drive components have an intrinsic self-braking action. Low pitch angle screws, high ratio speed reducers and worm gears are commonly used designs that inherently have self-braking features.

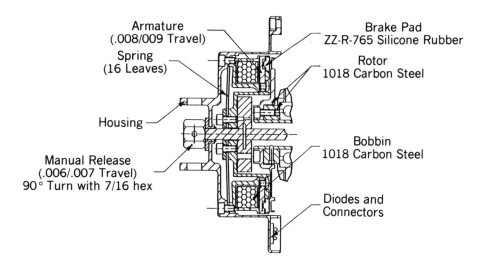

*Figure 2.25  Electric Brake*

## Industrial and Research Robot Arms

Two little known spray painting robots developed in the late 1930s are the first U.S. patents on the subject. The first "Position Controlling Apparatus" (Figure 2.26) filed in 1938 by Willard L. V. Pollard, is an interesting and unconventional mechanical design that features a flexible shoulder and wrist (Pollard 1942). Six axes of motion are powered by pneumatic cylinders and motors. A central arm pivots for shoulder pitch motion, and terminates in a U-joint that supports the lower arm. The lower arm is guided by two side mounted arms that swing side-to-side to control the arm in yaw motion (Figure 2.27). Three flexible rotary cables supply pitch, yaw, and roll to the wrist. A primitive electronic controller uses Thyrotron tubes to record position on a magnetic drum.

The second spray painting arm patent was assigned to the DeVilbiss Company, later to become one of the first industrial robot suppliers in the United States. Invented by Harold A. Roselund, the U.S. patent, "Means for Moving Spray Guns or Other Device Through Predetermined Paths" (Figure 2.28) was filed on August 17, 1939 and details a more compact design that is driven by a large multitrack drum cam (Roselund 1944). Valves actuated by the cam followers power the double acting hydraulic cylinders that drive the joints via worm gears and rack-and-pinion combinations. The wrist is surprisingly modern and is very similar to one of the GMF wrists in current use (Akeel 1989).

The extremely detailed drawings in both these patents suggest that prototypes may have been constructed. However, the inflexible controller (imagine having to change programs using giant metal drums!) was a major reason why robots did not come into general use until modern electronic controllers were available.

Another early patent that looks surprisingly modern was granted for a robot called "Improvements in or relating to Positioning or Manipulating Apparatus" invented by Cyril W. Kenward. The British patent was filed March 29, 1954 and was published August 21, 1957 (Kenward, 1957), and preceded George Devol's first robot patent by several months. It is an interesting parallel that Britain, birthplace of the machine tool industry, would also pioneer the idea of robotics.

Hydraulically powered, this dual arm, gantry mounted robot was years ahead of its time. The patent even speaks of robot self-replication. Featuring detachable grippers and gantry mounting, it could well be used as an illustration for a current research contract proposal (Figure 2.29). Another key feature of this design was complete internal porting of hydraulics and internal wiring, problems that went unaddressed in early hydraulic robots and often is not achieved in modern day hydraulic robots. Figure 2.30 shows the six-axes, hydraulically powered arm in cross section.

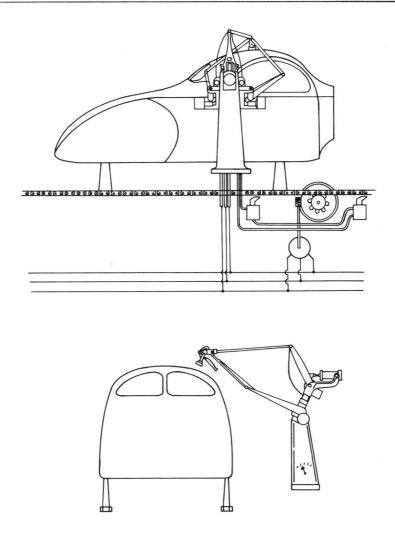

*Figure 2.26  Position Controlling Apparatus*

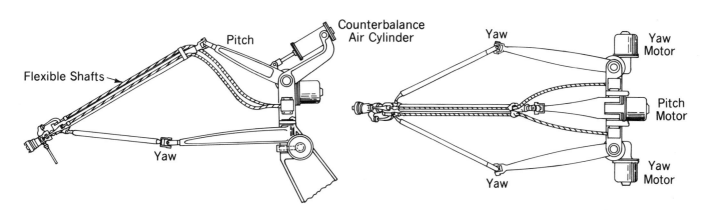

*Figure 2.27  Position Controlling Appartus: Detailed Views*

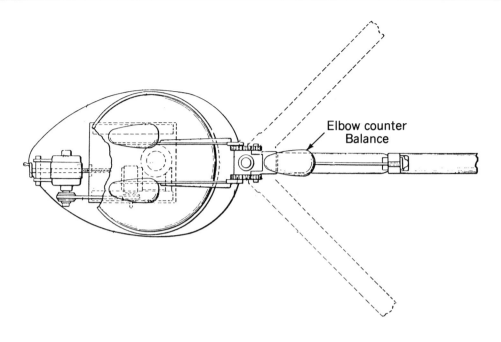

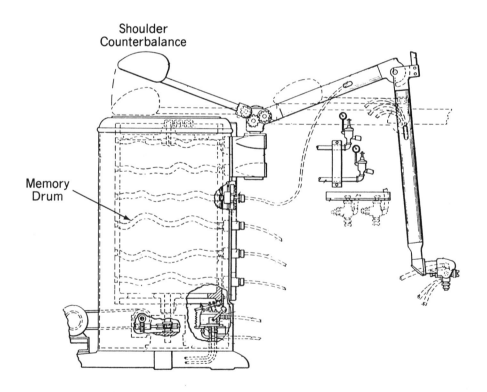

*Figure 2.28  Means for Moving Spray Guns*

Shoulder roll is powered by a vane actuator in the base. Shoulder pitch is produced by another vane actuator which is manifold mounted to the roll actuator. Shoulder yaw is accomplished in a similar way and shoulder roll is provided by an actuator in the upper arm. Elbow pitch and wrist roll also utilize vane actuators. In an odd coincidence, the popular Japanese robot toy "Armatron" by Tomy uses this exact kinematic configuration. Another interesting feature is the concept of quick change grippers, the first example employing this concept. The grippers also are hydraulically powered with internal porting. Features of this design are still found in modern hydraulic robots, such as the Sarcos and Schilling's manipulators.

The primitive control technology in existence was incapable of achieving what the mechanical designs were capable of performing. A poor match between design capabilities and control abilities simply did not allow complex and interactive assembly tasks to be performed. As with any new technology venture, skilled business talent also is required to effectively market the product. It was not until the Devol/Engelberger team addressed specific applications with a robot tailored to the existing mechanical and control technology limits that industrial robots graduated from the research lab to the factory floor.

A parallel development of robotic arms began at Argonne National Laboratory in the late 1940s under the leadership of Raymond C. Goertz, father of teleoperators. A classic design, they are still being manufactured by Central Research Labs, Red Wing, Minnesota. Although not computer controlled, these revolute coordinate type arms can be linked to modern robot arms.

The first of the master/slave manipulators to be widely used was the Model 4 (Figure 2.31). Developed in 1950, its load capacity was only 5 to 10 Lbs. Further improvements to the design included drive systems and wrists. The mechanical master-slave manipulator incorporated a parallel jaw-like gripper, roll-pitch-roll wrist, and pitch-yaw shoulder. Vertical motion also was designed into the arm. It was cleanly packaged; all drives are within the arm's tubular structure. Initially powered by the operator, electric versions were developed for more flexible installation. A windowed barrier, constructed of material suitable for the application, was placed between the master and slave arms. The window allowed the operator to view the material being manipulated while being protected from hazardous effects.

Another early nucleus of robotics research in the United States was General Electric. Led by Ralph Mosher, a team created an advanced teleoperator named "Handyman" in 1962. Extremely advanced for its time, it was used to demonstrate teleoperator control and was the first example of a three fingered hand with articulated digits. Another G.E. project at this time was Man-Mate. Each arm, basically a small crane, was controlled via an operator using early telerobot technology (Mosher 1971). This type

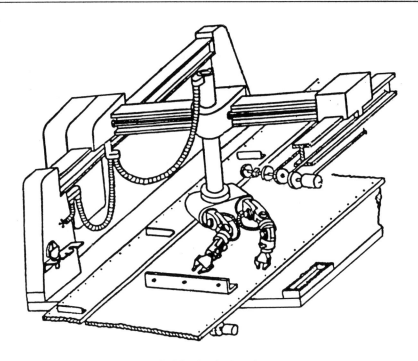

Figure 2.29  Manipulating Apparatus

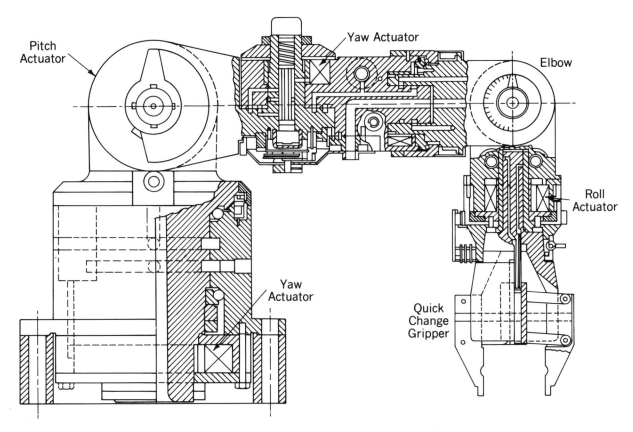

Figure 2.30  Cross Section of Manipulating Apparatus

of pioneering research is continuing in the United States largely for the handling of hazardous waste material.

The first commercial industrial robot arms were not the revolute arms developed by General Electric or Argonne, but cylindrical and polar coordinate arms developed by the Planet Corporation and American Machine Foundry (AMF). High load capacity and easily obtainable components dictated a cylindrical or polar coordinate configuration. The simpler kinematics these mechanisms provided could be matched more effectively to the primitive controllers of the time.

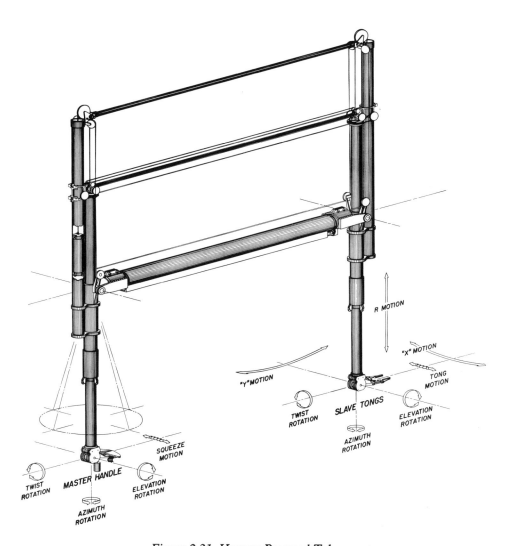

*Figure 2.31  Human Powered Teleoperator*

# Rectangular Coordinate

### Gantry Robot

A descendant of the overhead gantry crane, the rectangular coordinate or "gantry robot" has been with us since the 1950s. Developed originally for handling radioactive materials by the General Mills Corporation, now known as GCA Corp. Other applications include material handling, resistance welding, arc welding, and cutting operations with various tools.

The three main axes are x, y and z. Up to three degrees-of-freedom are used in the wrist, which is connected to the bottom of the z axis mast. This provides the endeffector with dexterity. The gantry provides the mobility of a man's legs. Several companies have also mounted entire six-axes arms to gantries to provide the best of both designs.

A three-axes gantry connected to a three-axes roll-pitch-roll wrist is depicted in Figure 2.32. Modularity is achieved by using various lengths of rail to customize the robot for any work envelope. The rectangular coordinate robot is basically a upside-down gantry with the x and y axis distributed in separate planes. Figure 2.33 shows a typical configuration.

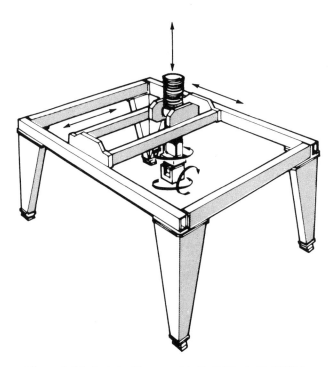

*Figure 2.32  3-Axis Gantry with Roll-Pitch-Roll Wrist*

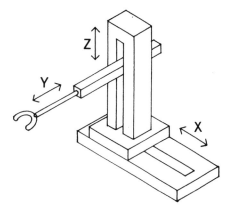

*Figure 2.33 Rectangular Coordinate Robot*

## Polar Coordinate

### Planetbot

The first polar coordinate commercial robot, called a Universal Transfer Device (UTD), was manufactured by the Planet Corporation during the tenure of president Frank Groeneveld Jr. (Figure 2.34). It was demonstrated at the St. Eriks International Trade Fair on Automation in Stockholm in 1957 under the brand name of Planetbot. It manipulated a baton through several obstacles and at a later date was used to dispense drinks, like several of its ancient forebears.

Drink dispensers have been a popular application of robots from ancient to modern times. It demonstrates man's endless quest for the perfect slave. Relatively simple to accomplish in a structured demonstration, robot bartenders bring the recipient into a fascinating interaction with the machine. Demonstrations involving the dispensing of drinks and maneuvering through a maze are classic promotional gimmicks for many industrial robot companies that wish to show the public a robot's capabilities.

At least eight Planetbots were eventually sold. Hydraulically powered, the machines had five axes of motion and were capable of 25 individual movements. First used in 1955 at the Harrison radiator manufacturing plant, a division of General Motors, Planetbot handled hot castings in several operations. The operations consisted of quenching, placement on a die press for trimming, and scrap disposal (Hole 1990). Originally it was controlled by a mechanical analog computer that resembled the margin setting mechanisms on an old mechanical typewriter. This laborious process, not far removed from the Scribe's cam system, contributed to its lack of commercial success. Originally it had two speeds, off and full; improvement came with the addition of a electrical sequencer. One of the quirks of this machine was that it would run wild in the morning because the hydraulic fluid temperature would be cooler than on the previous afternoon (this was before hydraulic fluid heaters were used in robots). The following description of the robot's operation comes from promotional material of the time.

> The Planetbot has five basic motions, which are: the grasping device on the horizontal arm opens and closes, and rotates 360 degrees. The arm will extend or retract, raise or lower, and rotate 270 degrees in a horizontal plane. In all, it is capable of performing a total of 25 individual movements during any one cycle or sequence and adjustments can be made in a matter of minutes to set the Planetbot for an entirely different cycle of operation. All movements are powered hydraulically either by cylinder or hydraulic motor and geartrains from a common hydraulic pump in conjunction with an accumulator. Hydraulic flow for each movement is controlled by direct acting double solenoid directional valves in conjunction with speed control valves. The main turret rotator has a double valving system for accelerating and decelerating.

It became clear that if a robot were to have a meaningful place on the production line, control technology needed to become faster, more flexible and responsive, and easier to program. Planet Corporation reentered the robotics field in 1981 with their hydraulically powered Planetbot. This large high load capacity robot, designed specifically for forging operations, has been a success.

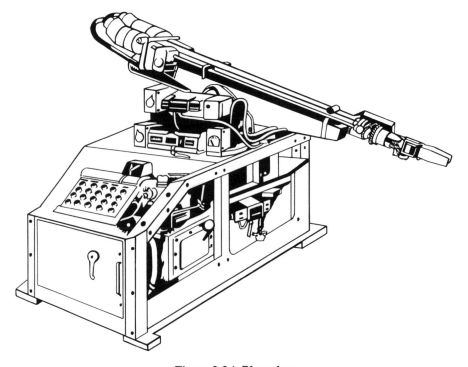

*Figure 2.34  Planetbot*

## Unimate

Although not the first commercially available robot as popularly believed, the Unimate made the greatest initial impact (Figure 2.35). Invented by George Devol, whose first robot patent application "Programmed Article Transfer" is dated December 10, 1954, it was a polar coordinate design that was successfully marketed with the help of promotional genius Joseph Engelberger (Saveriano 1981, Devol 1961). The first model was installed at General Motors' Ternstedt, New Jersey, plant in 1962 for automation of diecasting. Initial technical problems slowed its adoption, but by the mid-1960s it became an accepted part of assembly lines. Approximately 8500 of the units were sold, sixty percent of these to the auto industry. Load capacity is 300 Lbs in the Series 2000 and 450 Lbs in the Series 4000. The robots are not exactly lightweights themselves, weighing 2,796 and 5,000 Lbs respectively.

*Figure 2.35 Unimate*

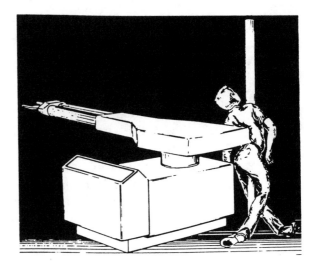

*Figure 2.36  Robot Inflicted Industrial Accident*

---

**Isaac Asimov's Three Laws of Robotics**

A robot may not injure a human being, nor through inaction allow a human being to come to harm.

A robot must obey the orders given it by human beings except where such orders would conflict with the First Law.

A robot must protect its own existence as long as such protection does not conflict with the First or Second Law.

Asimov 1983

---

Unimate industrial robots were an early violator of Isaac Asimov's First Law of Robotics (Asimov 1983), having been responsible for the death of several factory workers. One case involved a worker entering into a robot work space without following safety procedures. Ironically, he was pinned by the robot against a safety pole (Figure 2.36) designed to prevent the robot from overrotating (Collins 1985).

The polar coordinate design (Figure 2.37) was chosen for its high load capacity and simplicity (Dunne 1972). The system weighs in at 3500 Lbs. It is powered by an integral hydraulic pump which minimized flexible hoses, and could lift up to 150 Lbs with .050" repeatability. Shoulder yaw is achieved through a rack-and-pinion design driven by dual hydraulic cylinders. Shoulder pitch is powered by a double acting hydraulic cylinder that tilts the large aluminum casting forming the arm. A rotary manifold allows hydraulic fluid circulation to the arm and wrist cylinders (Figure

2.38). Wrists come in pitch-roll and pitch-yaw-roll configurations. The wrists are powered by hydraulic double acting cylinders, these cylinders power chain loops that engage sprockets which drive bevel gears. The latter produce the rotary motion in the wrist. The telescoping elbow motion for the arm is powered directly through a double acting hydraulic cylinder.

The use of readily available, stock components circumvented the need to develop actuators from scratch, but the price was a complicated and inelegant design. Nevertheless, they are still being manufactured, or rather remanufactured by Unimate's one-time competitor, Prab Robotics, Inc. of Kalamazoo, Michigan.

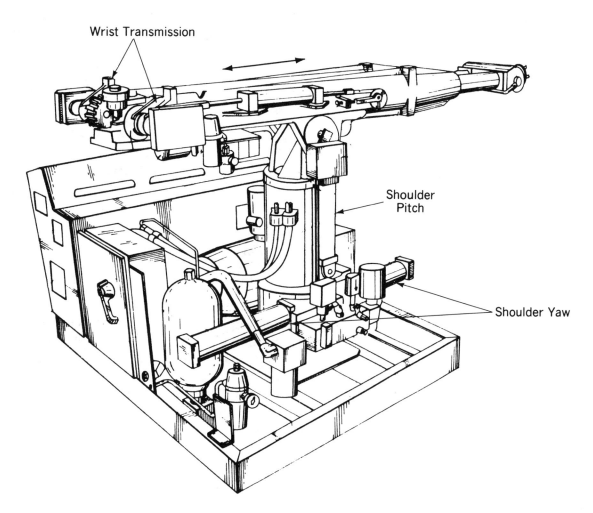

*Figure 2.37 Unimate Design*

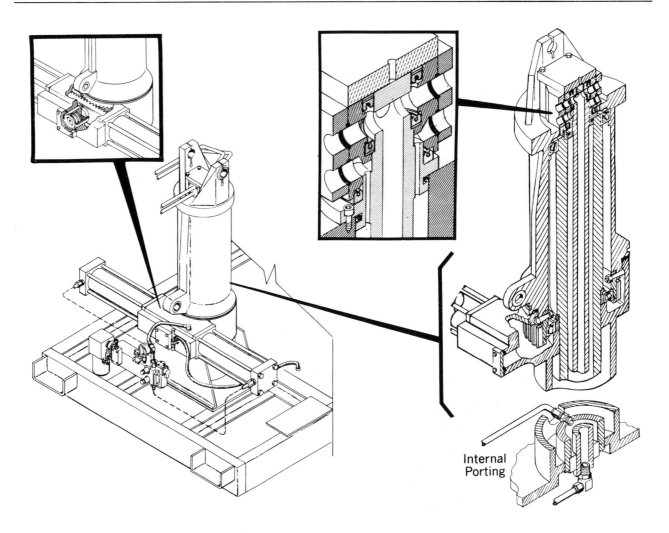

*Figure 2.38  Rotary Manifold of Unimate*

Internal
Porting

The proven design and high load capacity make this a viable product even today. The telescoping structure is ideal for reaching into punch presses, ovens, and other cavities where jointed, revolute arms have clearance problems at the elbow joint. The telescoping motion mimics the human elbow's ability to extend and retract the hand, but without the need to employ arcuate motion. The disadvantage of the telescoping arm is that it does not allow the arm to reach around obstacles, such as automobile door frames, which require planar reorientation of the endeffector. Incurable leakage problems created hazardous workplace conditions, in spite of the addition of a drip pan. Maintenance problems coupled with the inherently high maintenance costs of hydraulics, typically twice that of electrical drives, have caused this product to slip from its 1960 flagship rank in the American robotics industry (Naj 1990).

## Viking Arm

An interesting example of nontraditional, polar coordinate, robot arm design exists in the two Viking Landers (Figure 2.39) that were designed by Martin Marietta Astronautics of Denver, Colorado. The lander/orbiter spacecraft were launched on August 20 and September 9, 1975, from Kennedy Space Center for a 400 million mile journey to the planet Mars. The mission goal was to determine if there is any life on Mars. The Viking Arm, or surface sampler, was one of NASA's very successful early robotic devices. The harsh environmental conditions of vacuum and temperature extremes encountered in outer space had to be addressed in the design.

**Viking Lander**
1 S-band low-gain antenna
2 Facsimile camera No. 2
3 UHF antenna (relay)
4 X-ray fluorescence funnel and spoiler
5 Biology processor
6 Seismometer
7 S-band high-gain antenna (direct)
8 Wind and temperature sensors
9 Boom magnet cleaning brush
10 Gas chromatograph mass spectrometer processor
11 Descent engine (3)
12 Furlable boom
13 Collector head

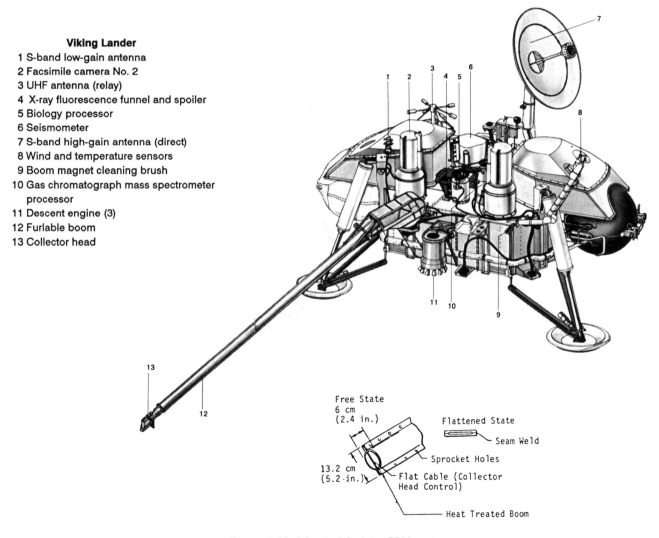

*Figure 2.39  Martin Marietta Viking Arm*

Viking employs a pitch-yaw shoulder and an extendable arm. Unfortunately this arm design has had little impact on industrial robots, but designers should take note of this creative design. In place of a conventional articulated arm or trombone type telescoping tube connected to a swivel base, extension of the arm was accomplished by two coiled ribbons that form a tube. The following is adapted from Mars Viking Surface Sampler Subsystem by Donald S. Crouch (Martin & Kuban 1985).

> The boom element is fabricated from two strips of Carpenter Custom 455 steel chemically milled to provide both a channel space for the integral flat cable, and a weight saving longitudinal taper along the length of the boom. Each strip is like a carpenter's tape measure, with the bow in two tapes facing each other forming a tubular space. Two halves of the boom are subsequently seam-welded together, sprocket holes are punched, and the material is heat-treated to the required "free-state" configuration. Following heat treatment, the flat cable is installed and the entire assembly is flattened and stored on the drum assembly. A series of rollers and guides flatten the free-state boom as it is rolled up on the drum. [The boom may extend up to 8 ft (2.4 m.).]

> Drive assemblies for the extend/retract, azimuth, and elevation axes are powered by hermetically sealed, solid-film lubricated, permanent magnet DC motors. The hermetic seal mechanical feed-through drive is accomplished by samarium cobalt drive motors that develop magnetic torque through a nonmagnetic membrane of the motor housing. A cermet element position potentiometer is coupled to each of the three drive assemblies. Sand and dust ingestion into the internal mechanisms is precluded by a silicone-coated Nomex cloth bellows surrounding the gimbal area, and spring-loaded Nomex cloth wipers that surround the flattened boom element located near the frame of the arm housing.

Construction materials for the boom unit included aluminum alloy, titanium and steel. Weight of the unit is 23 Lbs (10.4 Kg) and operating force is 30 Lbs (13.3 Kg).

The boom launch and landing restraint pins jammed during the shroud-eject sequence on one of the Viking Landers; the pin failing to drop free to the surface when the boom was extended. This was eventually diagnosed by laboratory modeling, which resulted in an action plan to free the pin by vibrating the collector head. The ability to have a robot remotely repair itself is significant, and argues against critics who believe that a robot is less reliable than a human.

# Cylindrical Coordinate

### Versitran

The first Versitran, the name a shortening of versatile transfer machine, (Figure 2.40) was manufactured by American Machine Foundry (AMF) Thermotool, Inc. The initial unit was shipped in 1960. Prab Robotics, Inc., then a maker of a downsized Unimate look alike, acquired the rights in 1978 primarily for the robot's electronic brain. This cylindrical coordinate machine has a rotary yaw shoulder, vertical travel for raising and lowering workpieces, a telescoping elbow, and a rudimentary wrist (Critchlow 1985). The hydraulically powered Versitran has a 100 Lbs load capacity and is still being manufactured today.

*Figure 2.40 Versitran*

The structure consists of a six-foot vertical column that is designed, like the Unimate, for 40,000 working hours. Covered with a sheet metal housing, the Versitran is also available with an accordion bellows (boot), which functions as a protective skin in hostile environments such as finishing applications.

The design has been improved upon by Cincinnati Milacron (now owned by ABB) in its model T3 300, which has a load capacity of 110 Lbs. The design patent on "Industrial Robot" was filed June 5, 1985 (Stackhouse 1987). The T3 300's more rational design approach is represented by a more rugged cast aluminum structure, that dispenses with the sheet metal cover, modern electric ballscrew drive and use of larger bearings and drive elements as shown in Figure 2.41.

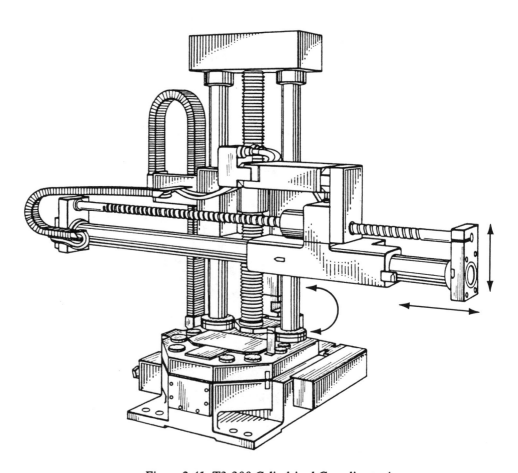

*Figure 2.41  T3-300 Cylindrical Coordinate Arm*

## Revolute Coordinate

### General Mills/Case Institute Arm

The first computer controlled revolute electric arm, which served as the prototype for countless future revolute arms, was developed by Norman F. Diedrich in the Digital Systems Laboratory at the Case Institute of Technology, Case Western Reserve University, Cleveland, Ohio. The arm demonstrated computer controlled manipulation by performing subroutines as specified by the operator (Johnsen and Corliss 1967). Support for this project came from the Space Nuclear Propulsion Office. Assembly of nuclear reactors for space applications was researched.

*Figure 2.42 Case Institute Arm*

Three goals for the arm were:

- Explore steps to improve man to machine communications for teleoperator systems.

- Maximize manipulator flexibility by reducing the number of operations required of the operator.

- Reduce operator fatigue.

Figure 2.42 shows a ceiling mounted application of the arm demonstrating the practical task of inserting pins in an array of holes. A surplus five degrees-of-freedom arm built by General Mills (at that time one of the leading manipulator builders) was modified to seven degrees-of-freedom and proportional control. Repeatability was 1/8". The arm's mechani-

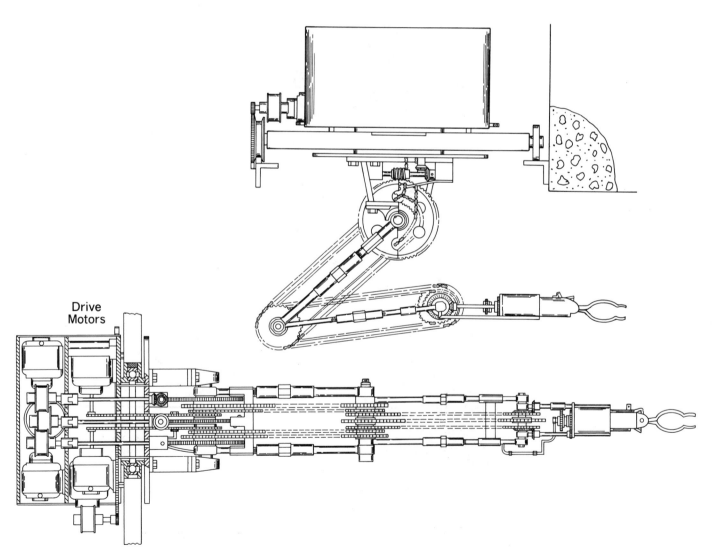

*Figure 2.43  Case Institute Arm Design*

cal design was invented by Arthur Youmans. The patent "Method and Apparatus for Performing Operations at a Remote Point" was filed on October 16, 1950 (Youmans 1958). Chain driven with all six motors located in the base, one-twentieth and one-tenth horsepower universal gear motors provide power. Through additional gearing (Figure 2.43) shoulder pitch and yaw (via roll), elbow and wrist pitch, wrist roll and gripper action are achieved.

## PUMA

Programmable Universal Manipulator for Assembly (PUMA), like the General Mills/Case Institute Arm, employs a revolute arm which has the appearance of the human arm. The revolute arm is the most popular robotic arm design, and has been used by many corporations. The main difference between revolute arms is the drivetrain. The PUMA (Figure 2.44) comes in three basic models: the 200, 500, and 700. The designers recognized that 90 percent of all components in a car weigh less then 5 Lbs (2.3 Kg), and PUMAs were commercially developed to address the perceived needs of the auto industry. The PUMA is a six-axes, revolute robot arm powered by direct current, brush type servomotors.

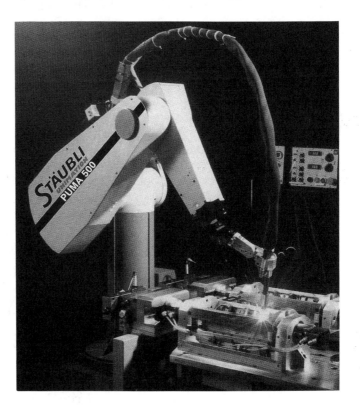

*Figure 2.44 PUMA*

Conceived in the late 1960s at Stanford University by Victor Shein-man, then a Stanford University graduate student, as part of an artificial intelligence project to create a microsurgery robot. Originally dubbed the Stanford Arm, several prototypes were sold in various anodized colors. It was commercialized by his company under the name VicArm. The first units used a telescoping elbow, but while a graduate student at MIT he developed the now familiar revolute configuration. An idiosyncrasy of these early units was motor to encoder connection via plastic spaghetti tubing, a less than reliable means of connection.

Licensed and reengineered by Unimation, Inc., the first unit was shipped on May 1, 1978 to the GM Technical Center. It is now marketed by Staubli of Switzerland. Staubli has made minor modifications including upgrading the wrist's bevel gears to spiral bevel gears. Shoulder yaw is powered by a motor in the base, its pinion driving a large bull gear for a 110-to-1 reduction (Figure 2.45). An antibacklash cartridge helps to control backlash. An inner column is supported by two bearings that are press fitted into the outer housing. This arm has the distinction of being the first to feature a box-like, monocoque housing that provides both maximum stiffness and lightweight. Speed reduction of all motors is accomplished through the respective drivetrains. Shoulder pitch is powered by the lower rear arm motor, which drives a small bevel gear meshing with a larger bevel gear (Figure 2.46). This second bevel gear ends in a pinion that drives the

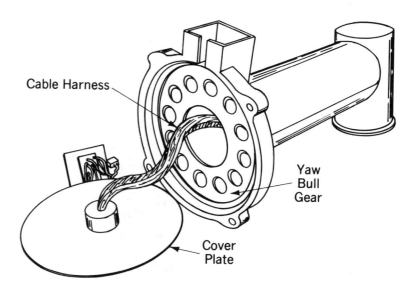

*Figure 2.45 Shoulder Yaw Drivetrain*

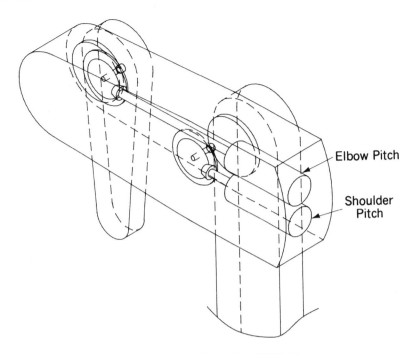

*Figure 2.46 Drivetrain of PUMA*

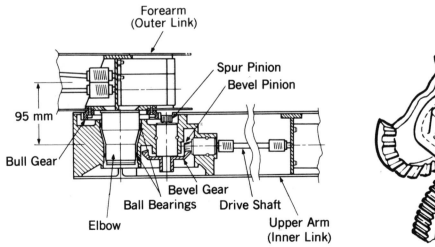

*Figure 2.47 Elbow Pitch Drivetrain*

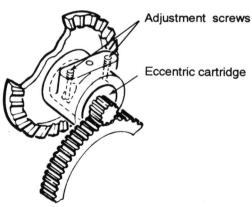

*Figure 2.48 Antibacklash Cartridge for Drivetrain*

large bull gear mounted on the upper arm housing. The two speed reductions in this drivetrain are valuable for increasing torque and precision. Taking advantage of gear reductions in this way is an important design decision when using gears. The shoulder is counterbalanced by the weight of the shoulder and elbow pitch motors that extend beyond the shoulder pitch axis. The bevel gears have caused the greatest problem associated with these arms (Figures 2.47 and 2.48). Bevel gear backlash adjustments in the arm and wrist reportedly have been misused, resulting in accelerated gear tooth wear.

Forearm pitch is driven in the same manner by the upper motor. The drive shaft extends the length of the forearm and terminates in a bevel gear that drives a larger bevel gear, which connects to a spur pinion driving a still larger bull gear. This type of drivetrain is emulated to some extent in the wrist. All the drive motors are enclosed within the housing which creates a very clean package with no external wiring. The wiring harness is routed through an upward central passage in the support column of the base. The wiring then passes through the center of the shoulder joint into the upper arm and is then routed through the elbow joint to the forearm.

Three motors in the forearm provide wrist roll, pitch, and tool roll. Helical flexible alignment units permit compact packaging by angling the wrist drive shafts. The flexible alignment also allows wrist roll by permitting the drive shafts to cross each other, like the radius and ulna in the human arm. However, undesirable compliance and loss of precision to the mechanical system of the arm and wrist degrades performance. The large number of couplings, gears and bearings required to produce motion results in a design that is relatively inefficient and inelegant. Design differences among the three models include relocation of a wrist motor in the 700, and variation of the forearm drive in the 200.

## Cincinnati Milacron

Cincinnati Milacron built large industrial robots primarily for the welding industry. The robot division has been purchased by Asea Brown Boveri (ABB), which continues to offer the product line. Cincinnati Milacron was one of the first companies to change from hydraulic to electric robots. Milacron pioneered the first computerized numerical control (CNC) robot, with improved wrists and the tool center point (TCP) concepts. The first hydraulic machine, the T3, was introduced in 1978, and closely resembled the General Electric Manmate, ITT Arm, and other predecessors (Sullivan 1971). Constructed of cast aluminum, it is available in two models of six-axes revolute jointed arms. The largest, the T3-776, uses ballscrew electric drives to power the shoulder and elbow pitch. The ball screws replaced the hydraulic cylinders originally used on the T3 robots (Figure 2.49). The elbow is a classic example of the intermediate drive elbow. The same techniques, only upside down, appear in the shoulder. Shoulder yaw is provided by the standard bull gear on a base mounted motor drive. End users have discovered that ballscrews are not sufficiently reliable and are pressuring for alternatives. The eventual disappearance of the ballscrew in industrial robots seems inevitable.

The drivetrain of the smaller T3-726 (Stackhouse 1983) represents an attempt at making an arm approximately the same size as the PUMA, but made more rugged by replacing the bevel gears with an all spur gear drivetrain (Figure 2.50). By virtue of this drivetrain, the T3-726 is the more interesting of the two shoulders. Payload is 14 Lbs (6 Kg) with a repeatability of ±0.004" (±0.001 mm). A motor in the base drives a three-stage

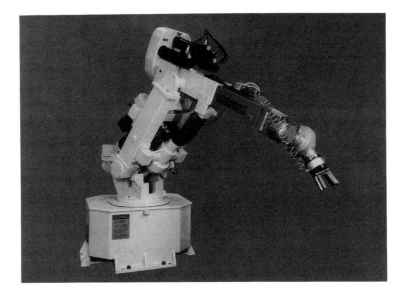

*Figure 2.49 T3-776 Six Axis Robot*

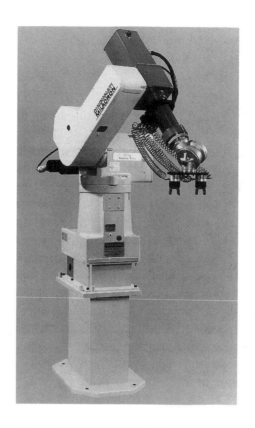

*Figure 2.50  T3-726 Six Axis Robot Arm*

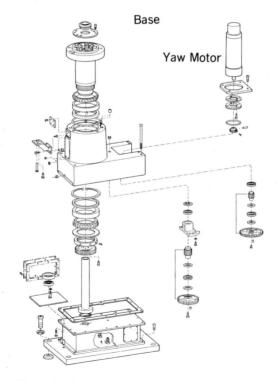

*Figure 2.51  T3-726 Shoulder Yaw Geartrain*

integral gearbox that rotates the shoulder for yaw. Timken bearings (Figure 2.51) are used throughout for ruggedness and precision. Shoulder pitch is provided by a horizontally powered motor that connects to a second integral three-stage gearbox (Figure 2.52). Another horizontal motor located near the shoulder motor powers the elbow by driving a six-stage row of spur gears that terminate in the elbow pivot (Figure 2.53). The forearm is a direct descendent of Hero's second century "barulkos", or weight puller, mentioned in his *Mechanica and Dioptra* (Figure 2.54) (Drachman 1963). The standard three-roll wrist is used. Like the Puma, all the drive elements are packaged within the robot's housing. The large number of parts in this arm and motor problems caused reliability problems which made this a less popular robot than other models. At last report, it is sold only in batches of ten.

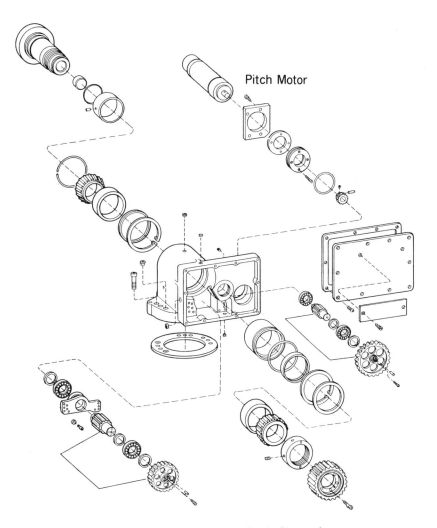

Pitch Motor

*Figure 2.52  T3-726 Shoulder Pitch Geartrain*

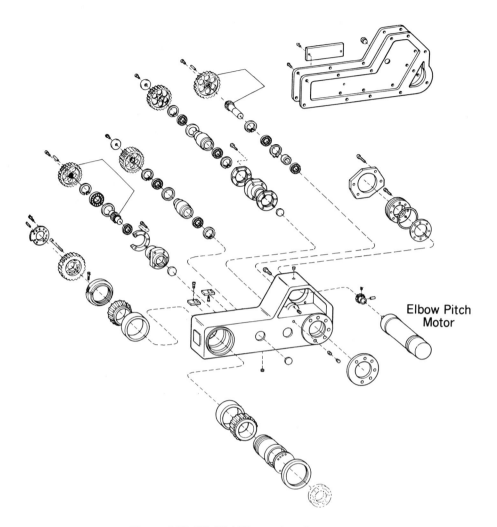

*Figure 2.53  T3-726 Upper Arm Drivetrain*

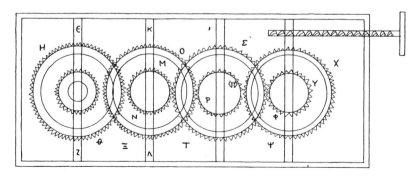

*Figure 2.54  Hero's 2nd Century "Barulkos"*

**GRI EDX-1**

Graco Robotics, Inc., (GRI) was launched in 1982 with technology developed by Ole Moloug, father of the spray finishing robot. ABB acquired GRI in 1991. Two U.S. patents have been granted on the EDX-1 design (Figure 2.55). The first, filed April 27, 1987, (Wood 1988) and the second, a design patent protecting the "look" of the design, was filed October 6, 1987 (Jones 1990). An outgrowth of their OM series of hydraulic robots, Moloug (1985), this electric arm borrows from many of the design concepts in the OM-5000. Roller chains and springloaded counterbalancing cylinders are two examples used throughout. Like the GMF P-100 the problem of mechanically backdriving requires a simple, low friction drivetrain that is backdriveable and counterbalanced (Figure

*Figure 2.55 EDX-1 Spray Finishing Robot*

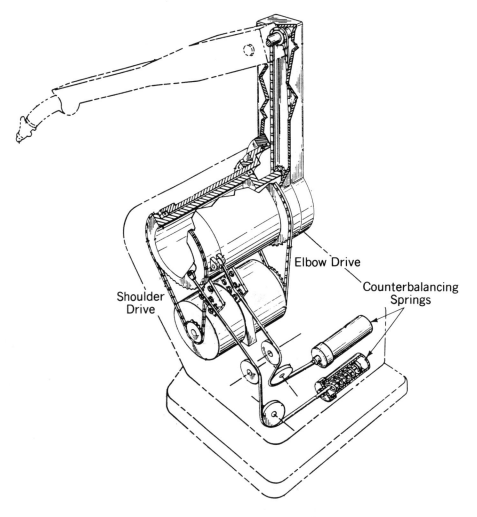

*Figure 2.56  EDX-1 Drivetrain (Wrist omitted)*

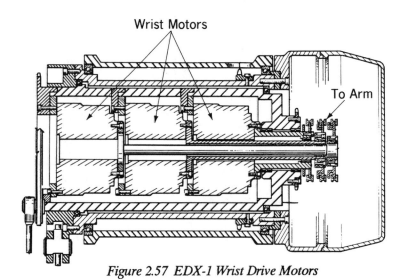

*Figure 2.57  EDX-1 Wrist Drive Motors*

2.56). This is achieved by placing the motors for the shoulder and elbow pitch in the base from where they drive, via chains and sprockets, two separate drums supported on large ring bearings. A second chain transfers motion through the upper arm to the elbow. The need to explosion-proof the arm by placing all motors in the base drove this requirement. Counterbalancing is achieved by springloaded cables attached to each drum and guided by a series of pulleys.

Unfortunately, like those of its predecessor the OM-5000, the chains of the EDX-1 are less reliable than linkrods or rotary shafts. Their main advantage is design flexibility and low initial cost. Chains not only provide speed reduction but also allow the sprocket centers to be located farther apart without a change in ratios. However, the initial cost break inevitably is reduced when the chains fail. The shoulder yaw motion is also driven by a chain and sprocket. Not surprisingly, customer complaints are forcing the adoption of more reliable drivetrains. The GMF P-100 has already adopted a more rational approach with linkrods and spur gears; belts were used for a time, but only as a means to transfer high RPM motion to the spur gears, and even these now are replaced by gears.

The EDX-1 is a good example of gradual rationalization of a robot arm design. Originally, three arrays of chains were used to transfer power to the Slim Wrist from horizontally stacked wrist motors (Figure 2.57). A triordinate system of shafts terminated in three pulleys that transmitted power through the upper and lower arm. The adoption of the Ross-Hime Designs' Omni-Wrist provided greater precision and reliability than its predecessor the Slim Wrist. With GRI's adoption of the Omni-Wrist, which uses rotary inputs, the pulley system was replaced by parallel rotary shafts that utilize miter gears at the shoulder and elbow. The relocation of wrist motors to the forearm is now almost universal.

**GMF P-100**

This is a six-axes, revolute, electric spray painting robot designed and built in the United States by General Motors Fanuc (GMF) (Figure 2.58). One version of the drivetrain is shown in Figure 2.59 (Bartlett 1988). The Japanese also have developed a similar design (Noguchi 1984). It represents a significant advancement over the complicated PUMA by virtue of its simplicity and more rational design. The arm may carry a 11 Lbs (5 Kg) payload. Shoulder pitch is powered by a motor, mounted on the shoulder housing, that drives a timing belt which drives a pinion/bull gear combination. The purpose of the timing belt is to provide clearance for the motor. This drivetrain provides a high speed reduction in a reliable, low friction

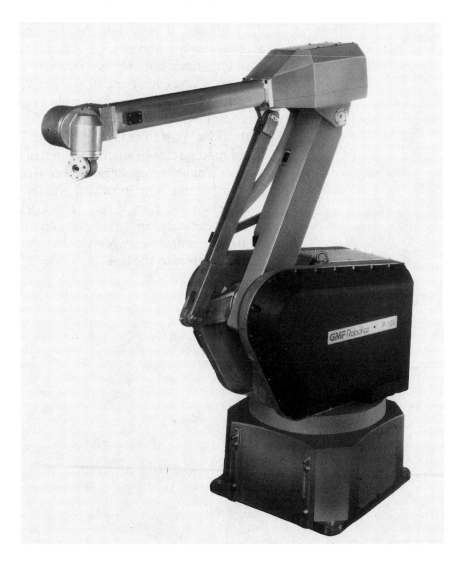

*Figure 2.58  GMF P-100 Robot*

manner. Upper arm pitch motion is powered with a drivetrain that is a mirror image of the shoulder's powertrain. Counterbalancing of the arm is accomplished by a tie rod that connects to a crank on the bull gear. The opposite end of the tie rod holds a spring under tension. A ball-and-socket joint permits the tie rod to pivot and the crank motion is generated by the bull gear. The low friction drivetrain with few mechanical connections makes it ideal for backdriving, which is needed for programming via lead through teaching. The wrist drivetrain consisted of pitch, yaw and roll motors mounted in the top of the lower arm driving rotary shafts to the bevel gear type wrist.

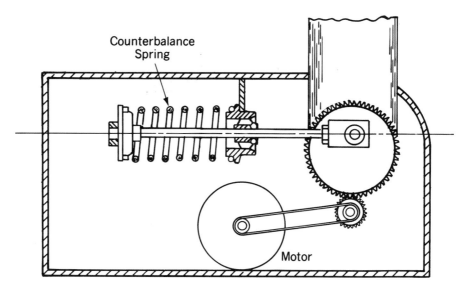

*Figure 2.59 Drivetrain of P-100*

## Parallelogram Drivetrain Arms

An additional linkage (Figure 2.60) was added to the basic revolute arm design in an effort to develop an arm with greater rigidity, and hence precision, than the T3. Although this makes for a bulkier profile, it does provide options in motor location because each of the pivots can have the motor driving it. Also, the design is more flexible and allows the use of linear or revolute type drivers. For practical load distribution purposes, motors are usually mounted on the lower two pivots, but the price is larger mass and expense. Range is also more limited than its simpler single bar cousin. The five bar is a simplified variation of the design.

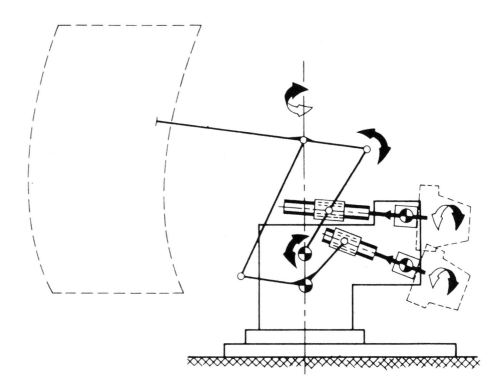

*Figure 2.60  Parallelogram Drivetrain Arm*

MIT has produced a horizontally oriented parallelogram arm that uses brushless, high torque, servo motors (Figure 2.61). The lower motor drives the foreground link and the upper motor drives the background link. The University of Santa Barbara has followed similar lines with their unusual arm (Figure 2.62) that uses an additional passive link to create wrist roll (Stoughton 1988).

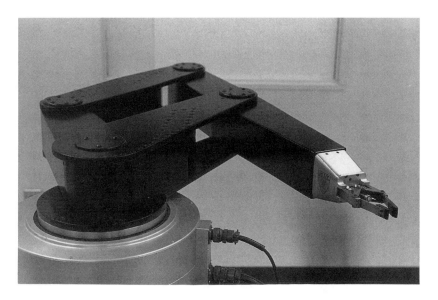

*Figure 2.61  MIT Parallelogram Arm*

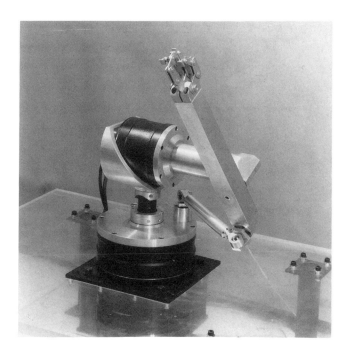

*Figure 2.62  University of Santa Barbara Arm*

## ABB Linkrod Design

Developed initially by ABB in the early 1970s, the linkrod drivetrain provides a very stiff design with great precision and control. Power is transmitted in both push and pull, in a manner that attempts to simulate the effect of human muscles and tendons. This design is not completely like the human body in that muscles can only pull. The human body requires that muscles and tendons work in a paired, but antagonistic relationship to achieve bidirectional motion. Two electric ball screw linear actuators drive the shoulder and elbow pitch axes. One problem with this type of design is that only 45 degrees of rotation can be achieved with a simple linkage driven crank. Unfortunately this lack of rotation and design flexibility is why so many manufacturers turned to chains and belts in the past. However, modifications to the joint's output via gears, as in the Dextrous Arm's shoulder joint, can provide the mobility and eliminate the need for chains or belts. Linkrod designs also suffer from coupling, occurring when the action of one link affects the action of another. Solid rods coupled by a few bearings have far fewer connections than chains and are more reliable than belts. They also provide relocation of the drive motors away from the joints that are being powered (Figure 2.63). Very reliable, they have been widely emulated by Thermwood, DeVilbiss, Cincinnati Milacron and others (Suica 1990) More and more robot manufacturers are being forced to this design by customers who have been displeased with chains, belts and elaborate geartrains.

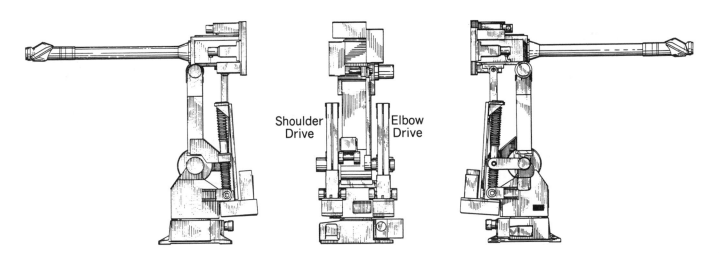

Shoulder Drive    Elbow Drive

*Figure 2.63 ABB Linkrod Design*

### Dainichi-Kiko Direct Drive Arm

An ongoing effort at MIT has been to design and build an improved arm that utilizes high torque brushless motors (Figure 2.64). (Read *Direct Drive Robots* by Asada and Youcep-Toumi for more information.) The basic concept was developed by Haruhiko Asada and Takeo Kanade in 1981 at Carnegie Mellon University, which produced the CMU D-D Arm Model 1 (Asada 1987). The U.S. patent was field September 30 1981 (Asada 1984). Subsequent models MIT DD-I and MIT DD-II were produced at MIT. The MIT DD-III arm has a velocity of 12 m/s with acceleration of over 5G and a bandwidth of 70 hertz. Features include decoupled dynamics, parallel drive with a six-bar link mechanism. One application envisioned is laser cutting. Brushless motors eliminate backlash from linkages and chains, as well as the friction caused by the necessary preloading of such systems. A basic revolute joint structure is used, with motors directly driving the joints (Figure 2.65). Shoulder yaw is achieved with a motor in the base. Two motors mounted on the shoulder

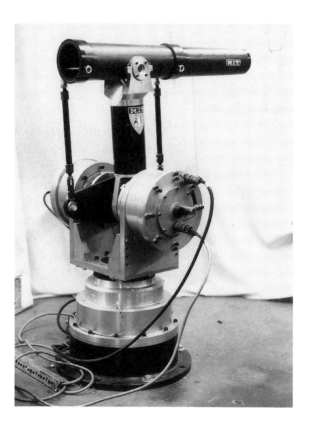

*Figure 2.64 MIT Arm*

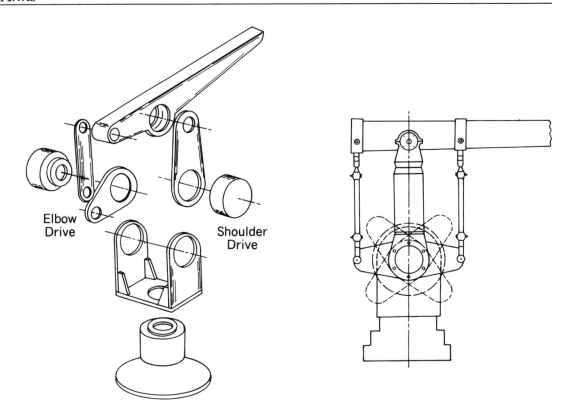

Elbow
Drive

Shoulder
Drive

*Figure 2.65  MIT Arm Drivetrain*

pitch axis provide up and down motion of the upper arm and the forearm. One motor connects to a double-ended link that connects with each end of the forearm. The second motor drives the upper arm in a simple direct drive connection. The main drawback of this system is the high cost of the direct drive motors, which also require braking mechanisms in case of power failure. This is offset somewhat by eliminating the need to counter-balance the arm. Graphite epoxy is used in the upper arm to reduce weight and therefore improve dynamic performance. The value of this approach is demonstrated by the Dainichi-Kiko, Inc. commercialization of the MIT Arm in 1986 (Figure 2.66). A drawback inherent to direct drive arms is both the wrist and shoulder have singularity caused by of the single axis nature of the components.

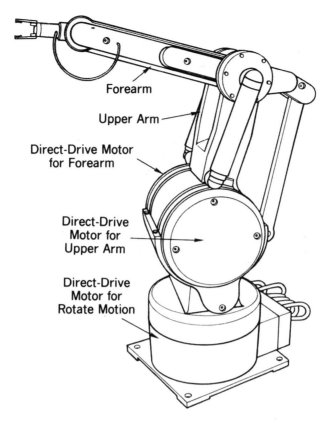

Forearm

Upper Arm

Direct-Drive Motor
for Forearm

Direct-Drive
Motor for
Upper Arm

Direct-Drive
Motor for
Rotate Motion

*Figure 2.66  Dainichi-Kiko, Inc. Direct Drive Arm*

## SCARA

Selective Compliance Assembly Robot Arm (SCARA) was developed by the Professor Hiroshi Makino at Yamanashi University Engineering Department in 1978-1979 and was first manufactured by Sankyo Seiki. It was reportedly inspired by the Japanese folding screen. Realizing that a large share of assembly operations require only a few simple motions, the Yamanashi team made a radical break with convention by developing robots designed to specialize and excel in a single assembly function rather than continue with the traditional quest to build a complex, general purpose structure. The manufacturing engineer breaks down the work process into several simple operations with perhaps some redesign of workpieces for ease of SCARA assembly. A robot is assigned to perform each task. This is a good example of successful robot design inspired by an actual application rather than designing a robot and hoping to find applications after the fact. Consequently SCARA robots are kinematically simpler and lower in cost, which permits high speed and precision on the order of .002" (.05 mm).

In most versions, distributed motors located in their respective joints provide power. In the Adept robot (Figure 2.67), two motors in the base power shoulder and elbow yaw motion. The elbow (Figure 2.68) is driven through two preloaded steel bands, thus eliminating the added mass of a motor at the elbow, which the shoulder would have to control (Carlisle 1987). Wrist motion is composed of a basic up and down motion powered by a ballscrew.

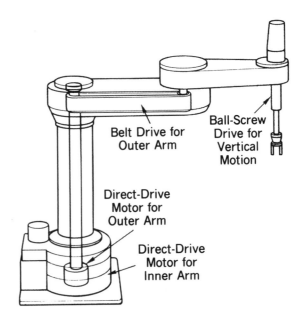

Belt Drive for
Outer Arm

Ball-Screw
Drive for
Vertical
Motion

Direct-Drive
Motor for
Outer Arm

Direct-Drive
Motor for
Inner Arm

*Figure 2.67  Selective Compliance Assembly Robot Arm*

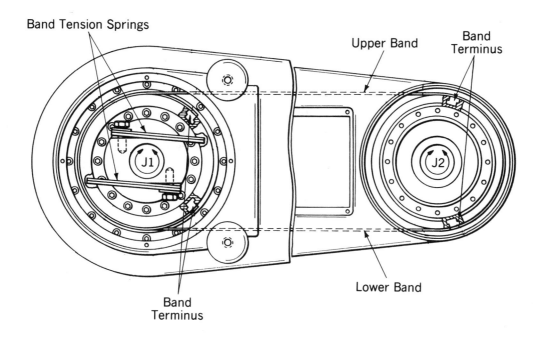

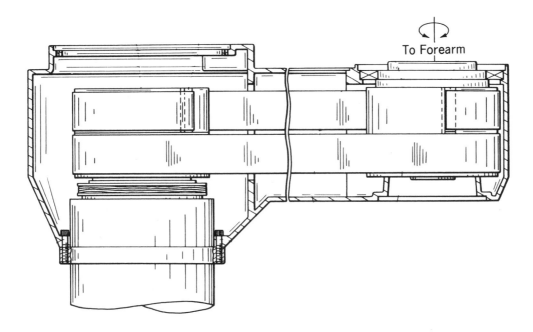

*Figure 2.68 Adept SCARA Design*

# Serpentine Arms

A growing need is perceived for ultrahigh dexterity arms for the inspection industry. The primary course of this evolution is anthropomorphic. Serpentine arms are a tangent to this, although their dexterity is greater. The ability to snake through convoluted passages lends itself to such applications as inspecting reactor cooling pipes, fuel tank baffles, and wing spars (Weiner 1989). Miniature versions have been incorporated into endoscopes for reaching into medical patients' bodies for tissue sampling and cauterizing, eliminating the need for large surgical openings. All of these designs consciously or unconsciously were based on the human spine. Although highly flexible, a penalty of poor precision is exacted as a consequence of the backlash generated by the large number of joints.

## The Human Spine

The multifunction spine (Figure 2.69) is a marvel of human engineering. It provides the human body with a flexible, muscle driven actuator for positioning the upper body, as well as a protective conduit for the spinal cord. Twenty-four vertebrae separated by intervertebral fibro cartilages (discs) make up the spine. The discs allow flexing of the spine by their capacity for compression in any axis. Each vertebra has protruding lugs (spinous process) for connection to the back muscles, that are arrayed along the entire length of the spine. Having the muscles located in this way simplifies the design by direct connection in the immediate area of the vertebra joint. Imagine how complex the tendon routing would be if every vertebra had tendons passing down the spine to a muscle lo-

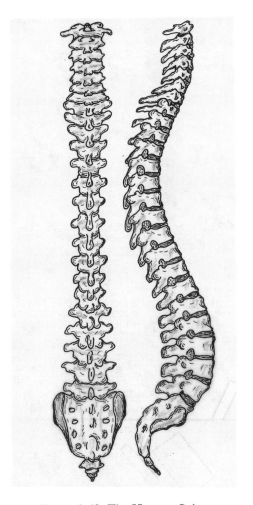

*Figure 2.69 The Human Spine*

cated at its base. The human back, already painfully strained and contorted into an S shape by walking erect, would be even more prone to failure as tendons stretch while lifting loads or to foreshortening from sitting long hours at a desk.

Figures 2.70 and 2.71 illustrate the spine's flexible kinematics and kinesiology. The human spine is capable of bending backward (hyperextension) 55 degrees. Forward motion (flexion) is 85 degrees. Side-to-side motion (lateral flexion) is a total of 120 degrees. Rotation is 45 degrees total. In addition the spine features a terminal pivot joint, which allows rotation of the head.

In general, it's not feasible to replicate the spine-like structures in inorganic materials owing to problems arising from cable management or from placing motors or hydraulic cylinders at the joints themselves. Multiple joint designs also produce more undesirable backlash then simpler structures (Miyake 1986).

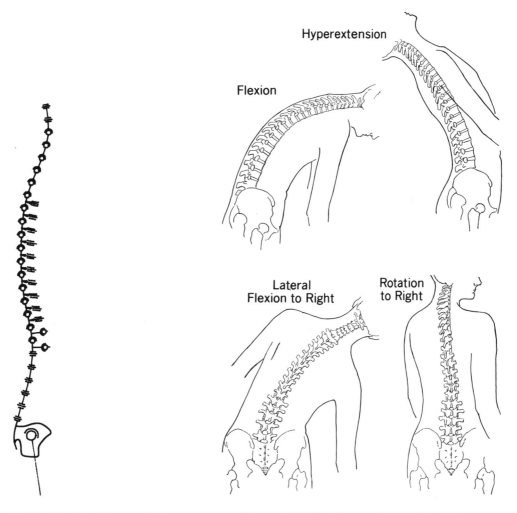

*Figure 2.70  Kinematic Model of the Human Spine*          *Figure 2.71  The Human Spine's Flexibility*

## Tensor Arm Manipulator

Tensor Arm Manipulator is one of the earliest attempts at a spine-like arm. The U.S. Navy has funded robotic arm research and development for over three decades because the average depth of the ocean is 12,000 feet, a depth at which human activity is impossible. Invented by V.C. Anderson et al., the patent was filed May 10, 1968 (Anderson 1970). A prototype was built at the Scripps Institute of Oceanography. Commonly called the Scripps Tensor Arm, it uses a series of U-joints attached to discs (Figure 2.72). Control is via a large number of nylon monofilaments. The number of filaments increase as joints are added, as each joint requires a separate set of control filaments (Figure 2.73). The passages that the filaments pass through are Teflon lined to decrease friction (Figure 2.74). Roll is provided by simply rotating the first U-joint and thusly each successive joint. Teflon ring bearings in each disc decouple roll from pitch and yaw.

The Scripps Tensor Arm is the first example of this decoupling technique, which previously was attributed to Moloug. Similar designs were patented by Olenick (1985) and Wilcock (1986).

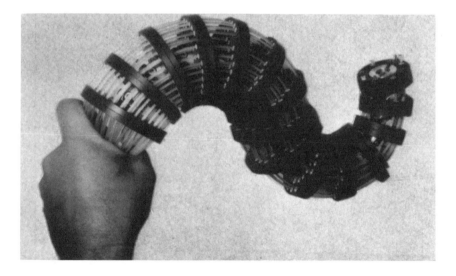

*Figure 2.72  The Tensor Arm*

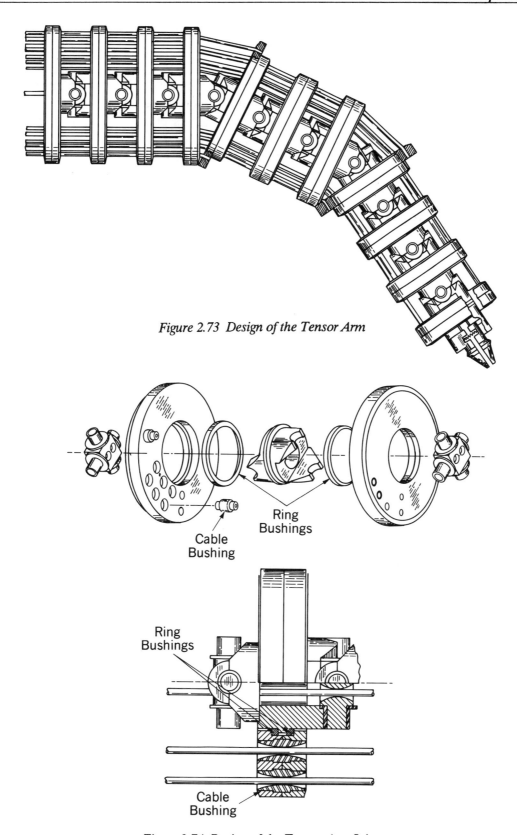

*Figure 2.73  Design of the Tensor Arm*

Cable
Bushing

Ring
Bushings

Ring
Bushings

Cable
Bushing

*Figure 2.74  Design of the Tensor Arm Joint*

## Articulating Mechanism

Articulating Mechanism (Figure 2.75) was invented by Ralph S. Mosher. The U.S. patent was filed September 24, 1969 (Mosher 1971). A clever, modular design, the arm is principally capable of one plane of motion spread over several centers. The plane may be rotated by rotating the base. Any of the sections can be rotated to permit a different plane of pitch motion, although once set this plane cannot be altered. A push/pull linkage drives the first joint, causing the first section to rotate about a pivot point. This motion is replicated by the joint above it via a diagonally positioned link. The link is connected to a pivot opposite the lower fulcrum point, transferring pitch motion to the joint and tilting it about its fulcrum ad infinitum. A bellows may be used to boot the arm for protection from hostile environments. SAC Taylor Hitec (Great Britain) has developed a version of this for servicing nuclear reactors (Kovan 1989).

The principal limitation of this design is the very poor precision inherent to all spine based systems. This is partly caused by the high ratio of input to output motion; one small error in the input produces a great error at the output. Also, each mechanical connection adds its own backlash to the system, and a large number of mechanical connections are required in these arms. Another limitation is that only one axis of motion is obtainable in this arm, which limits its use in applications that require greater flexibility. This limitation does, however, allow for a simpler design, lower backlash, and more reliable structure than designs that require two-axes joints, as shown below. One nice feature of this design is the modularity that produces a simpler, low cost concept. More complicated designs have been built to accomplish essentially the same prehensile motion. One example is the Arm for a Program-Controlled Manipulator described in *Robot Wrist Actuators* (Rosheim 1989b). Several uniquely shaped links are used in place of the articulating mechanism's modules.

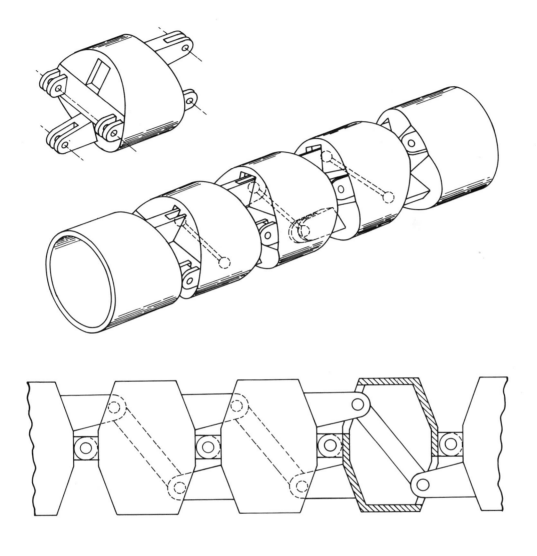

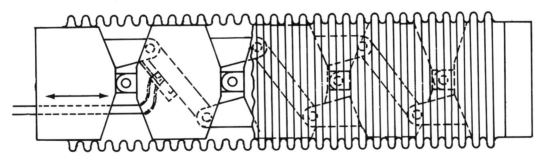

*Figure 2.75 Articulating Mechanism*

## The Space Crane

One of the early arm concepts for the United States Space Shuttle fleet was the Remote Control Manipulator for Zero Gravity Environment, also known as the "Space Crane" (Figure 2.76) The U.S. patent for this mechanism was filed September 18, 1970 by NASA engineer Frederick E. Wells (Wells, 1972) and constructed at NASA's Marshall Space Flight Center in Huntsville, Alabama. It is similar to the Ornithopter wing designs by Leonardo da Vinci described in Chapter 1. Each section, from one joint center to the next, measures 24". The six sections contribute to a twelve foot spherical working volume. A fabric stocking was used to protect the arm. In actual tests, conducted in a "high bay," the joints were suspended by a helium filled balloon to simulate zero gravity. The endeffector was able to place an object within 1/4" of the target. The precision was a surprising achievement, considering the high compliance caused by the cables, complexity, and length. Compliance is the flexing or distorting of a drivetrain component in which deflection is proportional to the resisting force. The arm logically stacks six hollow U-joints with cables fed to and through the joints for a high degree of pitch and yaw motion. In the drive box, two motors drive all the joints under independent control. One of the advantages foreseen by this redundant construction was that if a joint failed, the other joints could carry on the mission. Also, all electrical drives could be remotely located in the space craft, an important mainte-

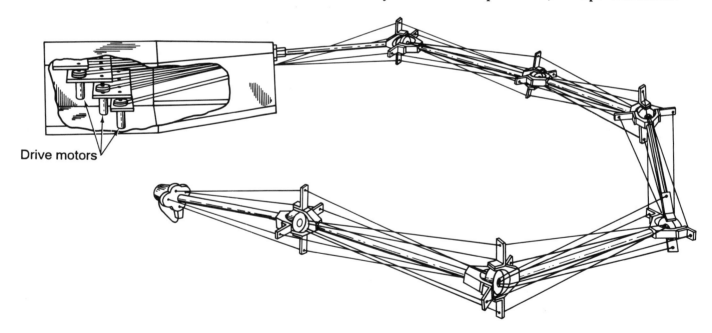

Drive motors

*Figure 2.76 NASA Space Crane*

nance advantage in space because of the harsh external environment. The cables are routed through the support tubes to each succeeding joint (Figure 2.77). As the first joint moves, each succeeding joint moves in an identical vector. Each motor is reversible and drives a drum that takes up and pays out at the same time. A simple, manually operated control system of two double-pole switches (one per motor) and two ammeters per joint comprises the electrical components. The ammeters tell the operator when something is sticking or if the rate of acceleration requested is too fast for the mass being handled.

Products of the flexible cable's inherent lack of stiffness are poor precision and maintenance problems. These problems must have been foreseen as the design was not pursued by NASA. Concerns about lubrication of the large number of joint bearings and cables in the cold vacuum of outer space also helped to prevent the design from being developed beyond the prototype stage.

This design is continually being reinvented. The Multiarticulated Manipulator (Iwatsuka 1986), developed primarily for visual inspections, has joints formed out of tubes that pivot on internal crosses.

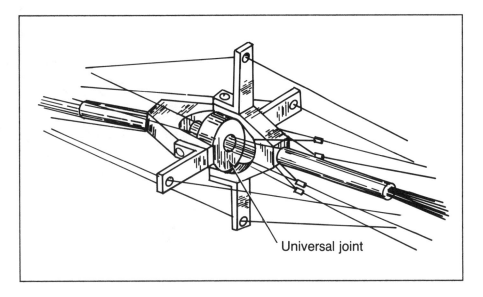

Universal joint

*Figure 2.77  Individual Joint of Space Crane*

**Serpentuator Manipulator**

This concept by H. Wuenecher, also of Marshall Space Flight Center, was developed at approximately the same time as the Remote Control Manipulator (Johnsen and Corliss 1967). Intended to aid astronauts by providing a long dextrous arm during EVA (extravehicular activities) (Figures 2.78 and 2.79), it was to be coiled in the shroud of a Saturn 5 rocket. It consists of links several feet long separated by joints and driven by electric motors. Eighteen links makeup the arm's sixty foot (18.29 m) length. A five-link prototype was constructed and tested with a tip force of 3 Lbs at simulated zero gravity. Like other early NASA robotic research projects the details of the Serpentuator Manipulator have been lost over the years.

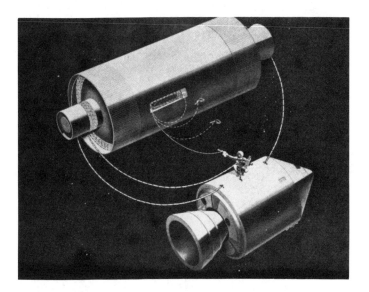

*Figure 2.78  Serpentuator Manipulator*

*Figure 2.79  Segment of Serpentuator Manipulator*

## Minsky Arm

Yet another attempt at high dexterity was created by Marvin Minsky, the artificial intelligence expert. Developed in 1968 at MIT, this electro-hydraulic, multijointed arm was built for the office of Naval Research for possible undersea work (Johnsen and Corliss 1967). Each of the twelve single degree-of-freedom joints has a position transducer parallel to the actuating cylinders, to provide feedback (Figures 2.80 and 2.81). Mounting the actuators near the joint makes this design among the closest to the human spine. The joints can be more precisely controlled when the

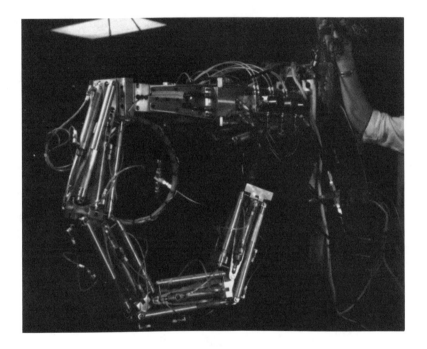

*Figure 2.80  Minsky Arm*

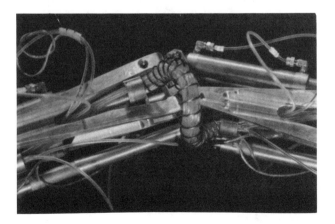

*Figure 2.81  Individual Joint of Minsky Arm*

drivetrain is a simple rod, which is clearly shown in the photographs, packaging is not as clean as in the human body. Hydraulic hose management was another problem which proved as serious as the filament and cable drives in the Scripps Tensor Arm and Space Crane. Multiple hydraulic hoses had to be routed from the source in the base to the hydraulic actuators near each joint. Control was by a PDP-6 computer. When wall mounted, it could lift the weight of a person. This design continues to resurface (Tesar 1989).

## The Spine Robot

The Swedish designed Spine robot (Figure 2.82) was an early attempt to compete at a high dexterity robot arm for spray painting automobiles. The Spine possesses four axes of motion in the arm itself and three in the wrist (Larson 1983,1985, Schreiber 1984 and Anon. 1985). Invented by Ove Larson and Charles Davidson, the U.S. patent was filed September 29, 1982. Larson's work continues with improvements (Bengtsson 1989).

It consists of a stack of specially shaped discs, called ovoids, controlled by opposing pairs of cables. These cables, under several tons of tension from the opposing pairs of hydraulic cylinders, run through stacked ovoids to form an antagonistic relationship. The base section contains the four hydraulic cylinders for the under arm, the valve block and filter, the position sensors for the under arm, and power sensors are mounted in the hydraulic oil distribution channels. The design was a commercial failure and has been discontinued. Maintenance problems must have been a contributing factor. Moreover, it was not so modular as bead-like. The ovoids, like beads on a string, were simply components that relied on outside forces (cables and the like) to make them function. Dependent on one localized power source, modularity was limited, although the ovoids are identical. If one cable is cut near the power source, its effect is felt throughout the system.

Many variations of this design have been produced including bellows that fulfill the pivotal function of the ovoids (Wilcock 1986), and multiple U-joints with driven pulleys (Wada 1987). Yet another variation places the bellows between plates that are fed pressurized fluid through a capillary system first developed by Victor Sheinman and Larry Leifer in 1965 (Sutherland 1990). This theme is continued further by placing miniature electric actuators to spread the plates apart (Ubhayaker 1990a and 1990b).

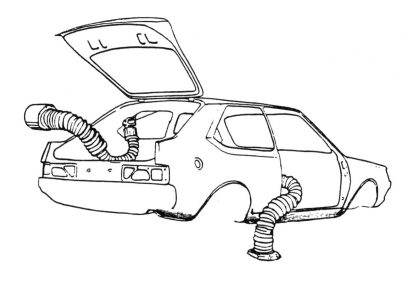

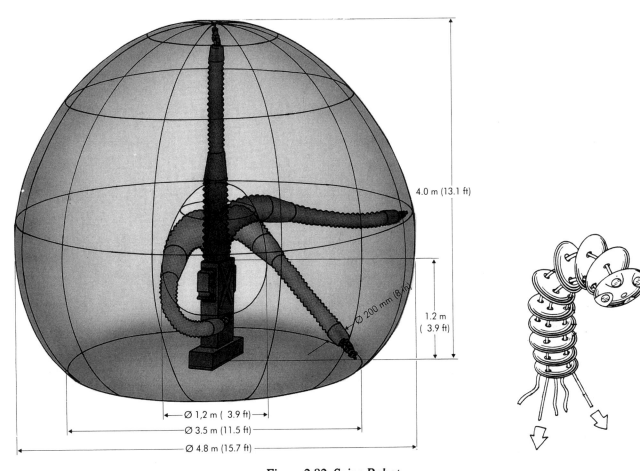

*Figure 2.82  Spine Robot*

## Expandable and Contractible Arms

Originality is shown in the Expandable and Contractible Arms. Invented by Motohiko Kumura, the European patent was filed August 22, 1984 (Hartly 1984, Kimura 1987). Like an accordion, three sets of linkages expand or contract to produce not only pitch and yaw but may also extend or retract the endeffector. Figure 2.83 displays the design. Miniature linear actuators, located between the links, expand or contract the individual elbow type joint. The links attach to the discs through U-joints that allow for decoupling of the pitch and yaw axis. Each unit includes at least three actuators. Both hydraulic and electric concepts are disclosed in the patent. One design disadvantage is the need for periodic replacement of the flexible bellows at each joint. The Expandable and Contractible Arms are marketed by Toshiba for nuclear power plant inspection in which the flexibility allows it to navigate tortuous paths.

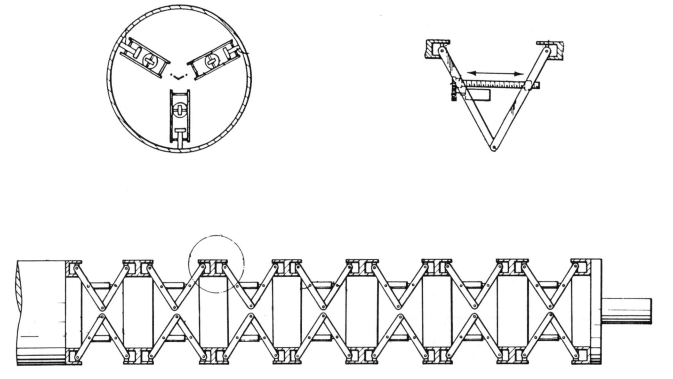

*Figure 2.83  Expandable and Contractible Arm*

## Modular Robotic Joint

The patent pending Modular Robotic Joint (MRJ) was invented by myself in August 1991 to overcome four of the key problems with serpentine arms: flexible bellows coverings, cable routing, modularity, and distributed actuators. The need for innovative, compound serial manipulators and robotic arms for manufacturing, or for EVA tasks such as manipulating tools and spacecraft modules, is evidenced by the previously mentioned designs. The MRJ satisfies the need for a wide range of motion and dexterity through simplicity and efficiency of design.

The enormous work envelope of three MRJ units configured as an arm is shown in Figure 2.84. Serpentine in its flexibility, it exceeds human dexterity and allows for access into confined spaces and through tortuous paths. Modularity increases the fault tolerance of the MRJ by providing redundant pitch and yaw axes in each joint. If one of the pitch or yaw axes fails, the opposite axis will provide mobility. If an entire MRJ unit fails, for example the shoulder unit, the elbow would be able to provide sufficient mobility to complete the mission. Although this violates one of my own rules about mounting actuators on the joint, the deviation from a design principle provides for modularity and lower cost.

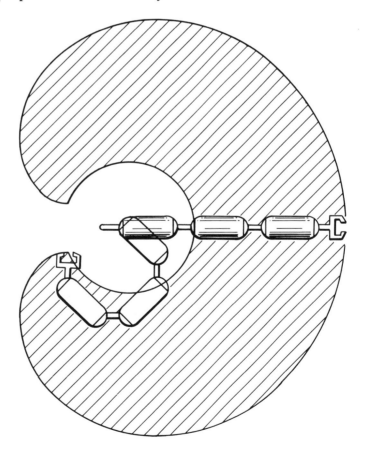

*Figure 2.84  Work Envelope of Modular Robotic Joint (MRJ)*

On a subsystem level modularity is achieved with the four identical linear actuators driving the joints. Minnacs produce high torque linear motion in a compact package. They are all identical and feature a unique zero backlash self-braking screw nut design. Self-braking is a valuable safety feature that prevents undesirable backdriving in case of power failure. MRJ features may be summarized as follows:

- Modular design allows reconfiguration into several limb types

- Singularity-free pitch-yaw-roll joints provide highest dexterity

- 2.5" through hole allows tangle free routing of cables

- Motion analogous to human limbs simplifies control

- Scalability from small to large diameters makes it adaptable to many tasks

- Linear actuator technology is nonbackdriveable and self-braking

- Solid metal housing provides ideal protection

Both joints consist of an inner universal joint supporting a double spherical shell driven by two Minnacs. The universal joints are supported by a hollow tube that permits passage of hoses, wires, or fiber optics. Reliability is increased by the joint's simple modular design and rugged packaging (Figure 2.85).

Each of the two MRJ joints has 180 degrees of singularity-free pitch-yaw motion about two axis centers (McKinney 1988). Four miniature electric linear actuators work in pairs to drive the double ball-and-socket like device. Two actuators per axis doubles the power available and provides redundant power to each axis. Each actuator is protected within its metal housing tube. The top and bottom of the housing has a lip with an O-ring that seals the mechanism from explosive gases, liquids and debris. The unit is also pressurized to further prevent external contaminants from penetrating the device.

The need for a flexible reconfigurable modular robot joint system can be readily met by this technology. Utilized as a shoulder, elbow and wrist, the MRJ may be used as an arm for, spray finishing, manipulating, cutting, welding, grinding, as well as cameras, x-ray, magnaflux, or ultrasonic inspection equipment (Department of Energy 1990).

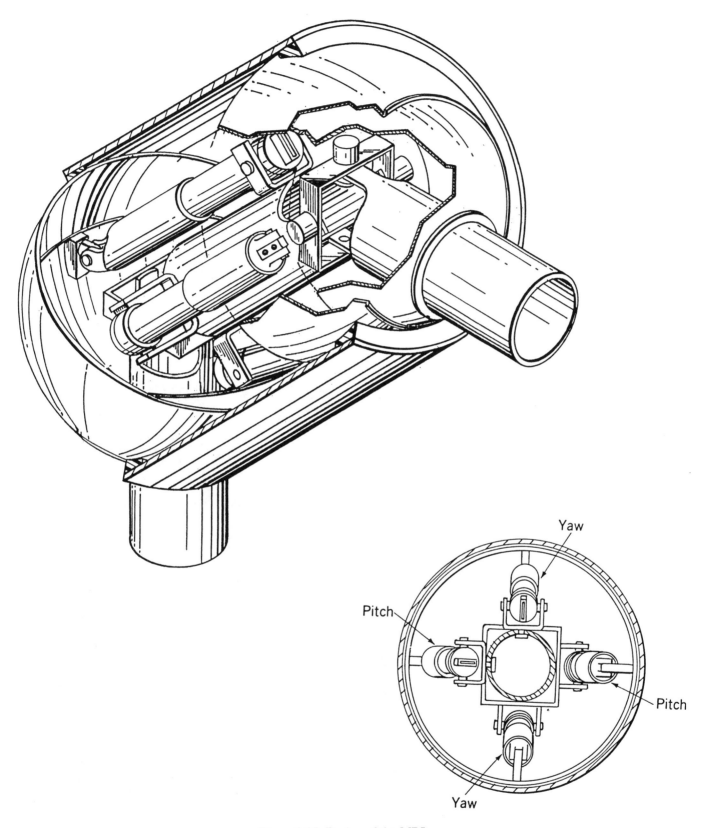

*Figure 2.85  Design of the MRJ*

## Hybrid Anthropomorphic

The hybrids were developed by robot companies that hoped to address a larger market than that found for conventional robot arms. They also wished to keep the reliability of the standard models while adding a new level of dexterity. This halfway approach has met with some success as evidenced by the designs below.

### Intelledex Model 660 Arm

The Intelledex 660 Arm has found application in **Class 10** clean rooms for handling electronic components. Invented by Allen J. Wright, the U.S. patent was filed January 11, 1985 (Wright 1985). Featuring .002" (.05 mm) repeatability and a 5 Lbs (2.3 Kg) load capacity, it is an early example of a robot with a more anthropomorphic shoulder (Figure 2.86) (Stauffer

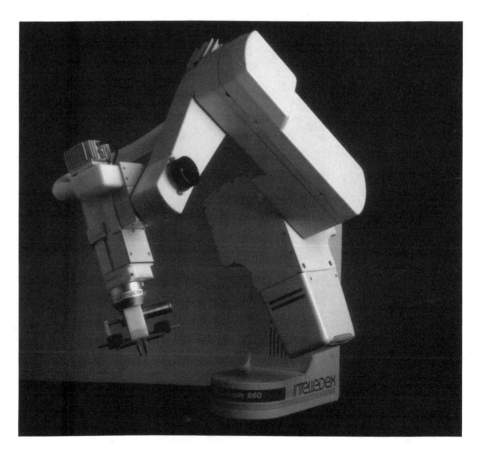

*Figure 2.86  Design of the Intelledex Model 660 Arm*

1983). Three axes intersect at a point to give the shoulder 180 degrees of pitch and 90 degrees of yaw. The roll motor is behind these actuators, probably for better weight distribution, but creates some roll axis motion problems due to its location (Figure 2.87). One problem is that the pitch and yaw joint, being a nonconstant velocity U-joint, requires a special compensation through programming to make the output velocity constant. The second problem is that the effective range of true roll without swinging the upper arm is limited to 90 degrees pitch and yaw, the maximum range of a single U-joint (Figure 2.88). Elbow pitch is powered by a motor through bevel gears. A simple direct drive approach is used for the wrist. The model 77 adds an additional axis to the base.

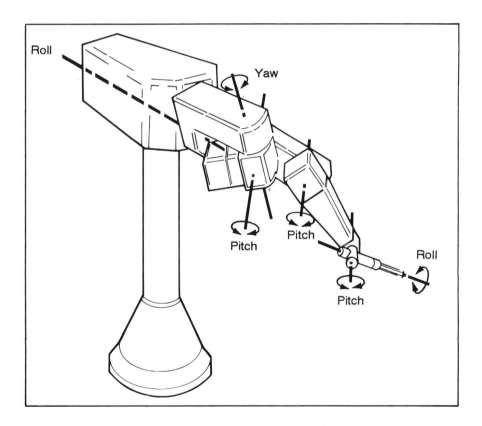

*Figure 2.87  Intelledex Model 660 Arm*

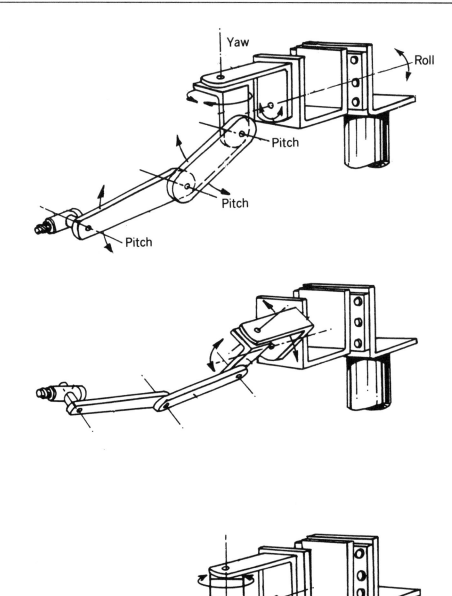

*Figure 2.88  Kinematics of Intelledex Model 660 Arm*

## ABB IRB 1000

The origin of the IRB 1000 may be traced back to the another Unimation innovation, the Apprentice (Figure 2.89), which was invented by Torsten H. Linbom. The U.S. patent was filed on "Portable Programmable Manipulator Apparatus" June 17, 1976 (Linbom 1978). Intended to be a portable welding apparatus, the device weighed only 125 Lbs (57 Kg) and could be positioned on top of a frame or box from which its gimbaled telescoping arm would weld seams. Programming was accomplished by the lead through method of teaching. Gimballing the telescoping arm provided the inherent counterbalancing needed for lead through teaching.

The six-axes IRB 1000 (Figure 2.90) is a more successful example of the Apprentice type design (Stauffer 1984). Introduced in 1988, it is claimed to be 50 percent faster and to use less floor space than conventional jointed revolute arms. Invented by Skoog et al., the first of a series

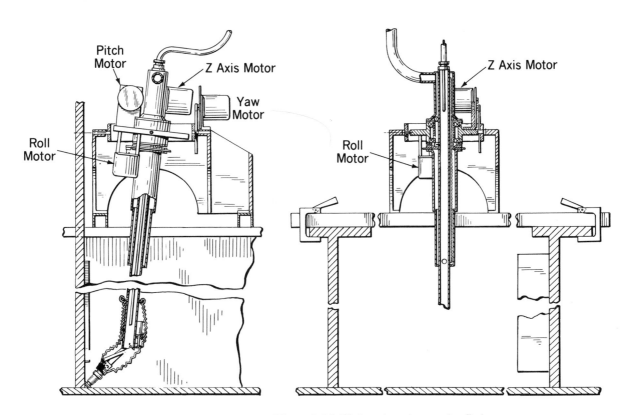

*Figure 2.89  Unimation Apprentice Robot*

of patents was filed Nov. 13, 1984 (Skoog 1987a, 1987b, Telld'en 1987). A 6.6 Lbs (3 Kg) capacity with .004" repeatability is featured. The IRB 1000 combines the telescoping elbow with the gimbal mount of the Apprentice in a large, servomotor driven, gimbal mount that surrounds a vertical telescoping arm. Although this design has six axes and appears to be very rugged and reliable, it uses a singularity ridden roll-pitch-roll wrist. Also, the disadvantage of suspending the arm in a gimbal creates a very large shoulder diameter and consequently bulky profile that is difficult to enclose. Telescoping the arm to achieve extension and retraction of the endeffector renders it unsuitable for reaching around corners or obstructions. For these reasons, this design results in a potential reduction of performance in more general applications.

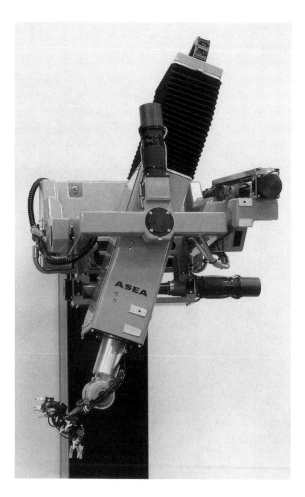

*Figure 2.90 ABB IRB 1000*

## Tetrabot

From England comes the Tetrabot (Figure 2.91) developed at the Marconi Research Center. Invented by Mark G. Vickers, the Tetrabot's U.S. patent was filed March 27, 1986 (Vickers 1988). It features a type of shoulder that is singularity-free within limits and attempts have been made to use this concept as a wrist (Rosheim 1989b). It is based on the Steward Platform, a telescoping, hydraulically powered tripod used in flight simulators to move the cockpit module. Its predecessor, a miniature version of Tetrabot, the "Gadfly" manipulator, was also designed at the Marconi Research Center. Several patents relate to similar designs (Neumann, 1988, Landsberger 1987, Simunovic 1986, Shum 1983, Shelef 1989). The central pole is gimbaled, and provides position control information. However, the joint cannot achieve 180 degrees of motion in the pitch and yaw axes because it has only one common axis center point, much like the IRB 1000. A positive feature is the tripod-like support that spreads over a large area and produces very high stiffness. Examples of applications for this include the Tetrabot from GEC Limited in Great Britain (Dwolatzky 1986).

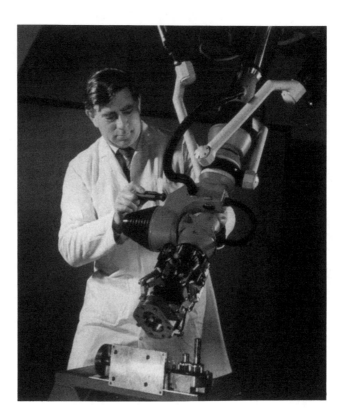

*Figure 2.91  Tetrabot*

The Tetrabot's load is concentrated in a central position. A 2 meter work envelope with a 22 Lbs (10 Kg) load capacity and .002" (.05 mm) repeatability is featured (Figure 2.92). Three linear actuators provide positioning of the roll-pitch-roll wrist in pitch and yaw as well as elbow extension. Use of three linear actuators to produce the two-axes motion gives this shoulder greater power in the ABB design, but it is bulkier as a consequence (Thornton 1986). Other tripod designs are covered in the various patents (Kochan 1991, Kohli & Sandor 1989, Neumann 1988, Shum 1983, Simunovic 1986, and Shelef 1989).

Revolute arms have the appearance of being anthropomorphic arms and are often labeled as such, but lack the implied performance capabilities. Shoulders and wrists made up of conventional, singularity ridden joints may have high precision, but suffer from poor dexterity and consequently have limited applications. Serpentine arms are the opposite extreme; their precision is poor because of high dexterity. Though very flexible, they are too complex and are less successful than the revolute arms. Between these two extremes is the Dextrous Arm, combining the ruggedness of the revolute with the dexterity attributes of the serpentine. A design concept proven through countless human beings, this hybrid approach is crucial to the robot of the future. Commercial applications include robot arms for welding, spray painting, and applying sealant and adhesive compounds to automobiles, where there are many contours and compound curved surfaces. Inspection in confined, hard to reach spaces, such as aircraft fuel tanks and wing assemblies, also demand the flexibility, strength, and precision of an anthropomorphic arm (Weiner 1989). Space station construction, satellite servicing, planetary exploration, and nuclear handling all require dextrous arms. Undersea exploration and salvage, research, as well as transoceanic cable repair demand a man equivalent robot arm capable of operating over long periods in a hostile environment. In the ensuing section we will discuss robot arms which emulate human capabilities, culminating with a design for a new, state of the art, seven degrees-of-freedom, man equivalent arm.

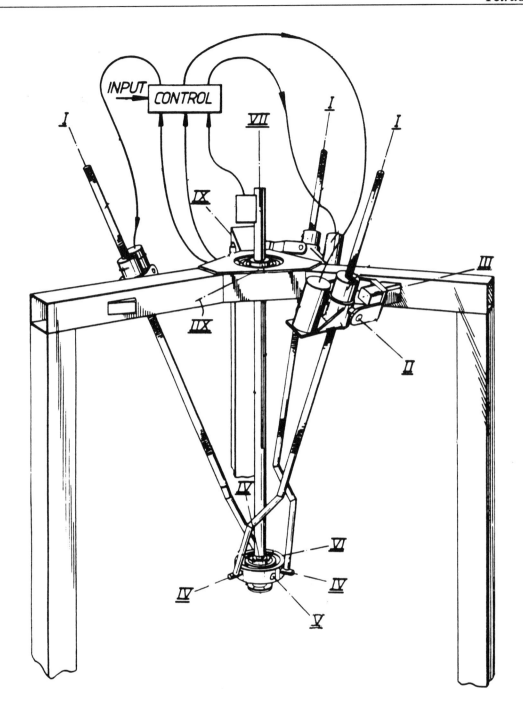

*Figure 2.92 Design of Tetrabot*

## Anthropomorphic Arm Designs

Throughout this Chapter I have shown the capabilities of the human arm and attempts at replicating it by mechanical means. A summary of the arm's kinematics in a functional manner is provided below. The wrist and hand are covered in detail in the next two chapters.

The arm, kinematically shown in Figure 2.93, consists of three joints (shoulder, elbow, wrist) plus a highly differentiated endeffector (hand). Each joint provides different types of motion, and accomplishes different functions. The shoulder provides gross singularity-free movement (240° pitch, 180° yaw, and 90° roll) and requires greater stiffness for the high loads it carries. The elbow produces intermediary motion, bringing the

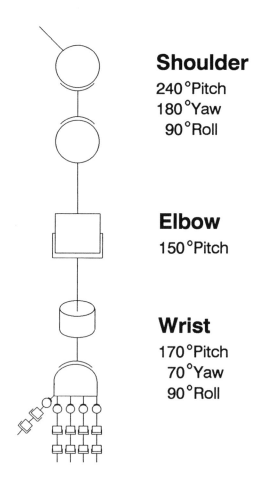

**Shoulder**
240°Pitch
180°Yaw
90°Roll

**Elbow**
150°Pitch

**Wrist**
170°Pitch
70°Yaw
90°Roll

*Figure 2.93 Kinematics of Human Arm*

endeffector (and its burden) closer to and farther from the torso, and provides a finer 150° degree of pitch motion that complements the shoulder's movement. The elbow also adjusts the locating of the endeffector with respect to a plane or angle set by the shoulder. This makes possible the use of the forearms as crude endeffectors used for bracing, or permits reaching around objects. The wrist provides ever finer positional control and produces small, singularity-free movements that complement the shoulder and elbow with an additional 170° pitch, 70° yaw and 90° roll capability to the endeffector.

Each joint provides a different function:

- Shoulder—gross pitch, yaw, roll with high stiffness
- Elbow—extend/retract, reach around
- Wrist—fine pitch, yaw, roll

Together they produce a system with seven-degrees of motion, and the hand adds an additional manipulative capacity. Changing one joint is to reduce and alter the ability of the system to function according to plan.

The following designs show a progression from simple revolute joints to more complex, and more anthromorphic joints. Gimbals and ball-and-socket joints are used to replicate the dextrous and fluid motion of the human joint. The more fluidity in the design's movements, the more versatile and broader its application.

## Robotics Research Arm

Development of this arm started in 1978 with a goal of designing a high performance modular manipulator for advanced applications in the space and defense sectors (Figures 2.94). An arthropod construction is applied, which makes for high stiffness but less than ideal dexterity. Harmonic drive motor modules with brakes are used throughout, serving as either roll or pitch joints. The first patent was applied for on February 14, 1989 (Karlen 1989, 1990). The design of the modules was a logical choice, as it used existing technology of harmonic drives and well established, albeit singularity-ridden kinematics (Figure 2.95). The arm has an advantageous payload-to-arm ratio because of the lightweight tubular structure and high performance motor/speed reducer modules. Internal pressurization against environmental contamination is achieved with its all metal revolute joints. The three sizes vary in length from 47" to 83" long. Repeatability of the largest size K-2l07HR is .0005". High payload-to-arm weight ratios are from 5:1 to 10: and load capacities from 20 to 50 Lbs. The shoulder, elbow, and wrist are kinematically identical.

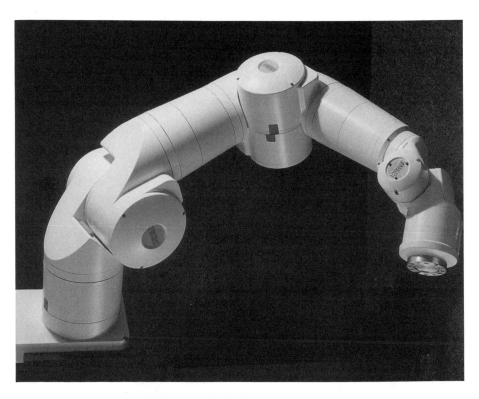

*Figure 2.94  Robotics Research Arm*

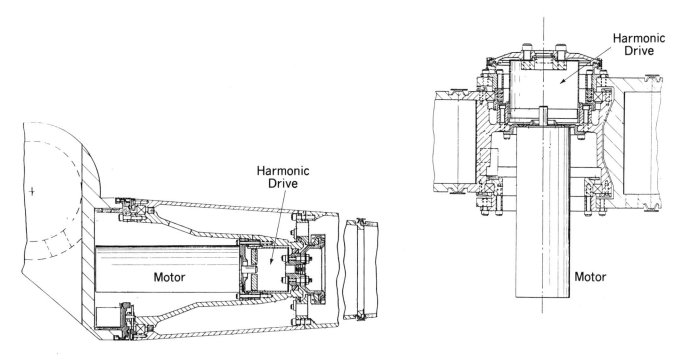

*Figure 2.95  Design of Robotics Research Arm Modules*

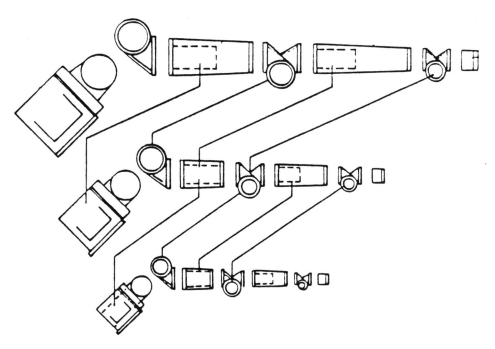

*Figure 2.96  Robotics Research Module Family*

Modularity is a a key feature of this design (Figure 2.95). The roll and pitch joints are not only identical in design but also are used throughout the different sized arms. For example, the shoulder of a smaller arm may become the wrist of a larger arm. Such interchangeability of parts provides numerous economies for both the manufacturer and the customer through more profitable production runs, smaller parts inventories, shorter logistics delay times, simplified service and repair. However, this concept has been taken too far in this design. Although this is a seven-axes arm, all joints are configured like the elbow, apparently as a design expedient. Transverse mounting of the pitch motors creates a bulky profile, as opposed to in-line orientation. These joints are a step backwards in design technology, as the shoulder, elbow and wrist by definition are kinematically and structurally unique mechanisms. The shoulder and wrist need greater dexterity than the elbow. While some compromise is inevitable, it should not interfere with overall performance. In the Robotics Research Arm dexterity has been sacrificed for the sake of modularity, load capacity and precision. The use of single revolute joints in this design results in the Robotics Research Arm achieving high precision and load capacity within the narrow limits of the joint's dexterity. The three joints' high precision and load capacity is of little use if you are unable to manipulate the workpiece because of singularities in the shoulder and wrist.

## LTM

Laboratory Telerobot Manipulator (LTM), originally named Man Equivalent Telerobot (METR), started as a modest $600,000 to $800,000 project but soon blossomed into a $3,000,000 expenditure (personal communication with A. Meintel 1991). The U.S. patent was filed April 21, 1987 (Martin 1989). It began as an attempt at designing high dexterity robot arms utilizing pitch-yaw-roll joints (Kuban & Perkins 1989). The METR was initiated at Oak Ridge National Laboratory under William R. Hamel for Langley Research Center (Figure 2.97) as an outgrowth of the Advanced Servo Manipulator (ASM) (Holt 1988). Two goals for this project were to reduce the very high levels of backlash by introduction of traction drive technology and to manipulate a 25 Lbs (11.4 Kg) load. Three similar traction drive joints with motors located in the arm sections and joint terminus make up the modular arm. The design (Figure 2.98) achieves an increased range of motion in the pitch axis by using a U-joint from which the top fork of the clevis joint has been removed. Pitch and yaw motion are powered by two motors located in the forearm of the wrist. The output shafts of the motors drive the wrist through a differential traction cone arrangement, providing 180 degrees of pitch and 90 degrees of yaw about the intersection point of the pitch and yaw axes. The joint is singularity-free, if operated within its boundaries. It should be noted, however, that unlike the human wrist and shoulder, the LTM's pitch motion is biased to one side because of the wrist's modified U-joint structure. An integral four-to-one reduction in the traction cone drivetrain is valuable, because it gears down the forearm motors, and increases wrist torque and resolution. Continuous bidirectional roll motion is provided by an additional motor located in the output arm (Kuban and Williams 1990) (Martin 1989).

One major disadvantage of the wrist is the mounting of the roll motor at the end of the wrist. This is unavoidable in this design because of the need to maintain cost effectiveness and precision. However, the added weight of the roll motor decreases the load capacity of the wrist, makes the wrist bulkier, and increases the distance between the endeffector and the wrist's center of rotation. The latter effectively increases the rotational inertia about the center point, resulting in a loss of wrist precision. The location of the motor compromises the modularity of the wrist because the motors are not centrally located. Power for the roll motor and position feedback information must be transmitted via flexible wires encased in twist capsules. A twist capsule is an encapsulated coil of wire similar to a spring that allows electrical circuit(s) to be maintained within a limited rotation about an axis. However, the capsules introduce another potential failure item and additional maintenance problems.

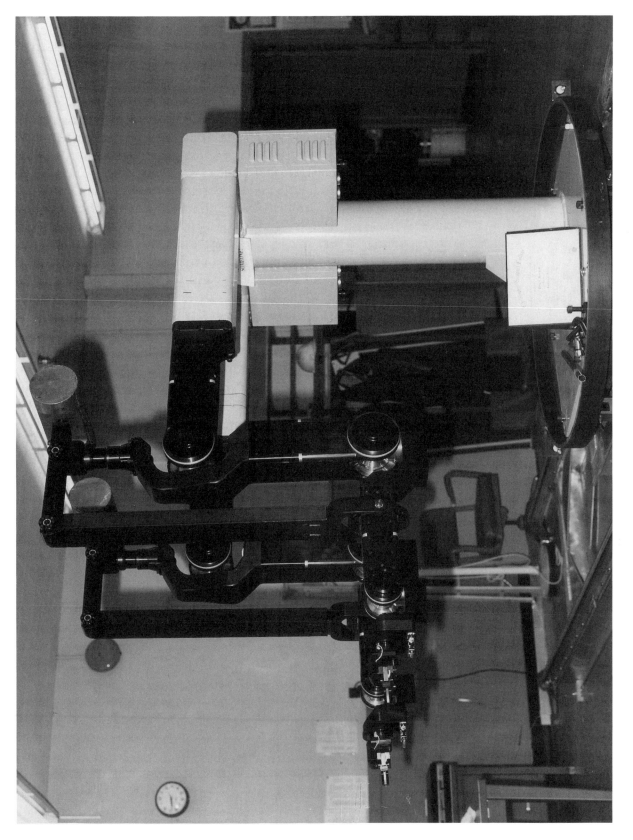

Figure 2.97 Laboratory Telerobot Manipulator (LTM)

The LTM's drive system is unusual in using a traction cone drivetrain, usually used in high speed, infinitely variable speed transmissions (Kuban and Williams 1990). The drive was chosen over high tolerance bevel gears to reduce the backlash and friction that inhibit precision and backdriving. However, the traction cone drivetrain is mechanically less efficient and more expensive (the cones are literally gold plated) than other, more established and simpler drive systems. The extremely high preloads caused structural failures during development. Ironically, conventional speed reducers with geartrains are used to power the pitch and yaw axes actuators so that backlash and compliance is introduced at the input to the system. Regardless of how well the traction drives controlled backlash, it was introduced into the system by the motor gearboxes. The roll axis is entirely gear driven and suffers similar backlash problems. Shoulder range of motion is less than one-half of the "man equivalent" in either pitch or yaw. The problems of implementing the new traction drive technology to fix a weak initial concept proved to be a fatal flaw that led to cascading problems and finally proved to be insurmountable.

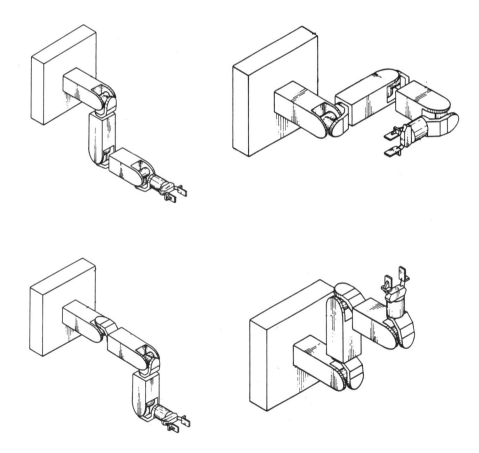

*Figure 2.98  LTM in Various Poses*

**FTS**

The Flight Telerobotic Servicer (Figures 2.99 and 2.100 is another seven-axes arm developed by Martin Marietta for Goddard Space Flight Center space applications. This $200,000,000. expenditure was an outgrowth of the Proto-Flight Manipulator Arm (PFMA) developed in 1977 and delivered to Marshall Space Flight Center (Brunson 1986). The PFMA arm consists of a three-axes shoulder joint and a single-axis elbow. Martin's robotic heritage extends further back to the 1963 Space Operations Simulator, which includes a six degrees-of-freedom moving base carriage and a seven degrees-of-freedom slave manipulator arm. A dual path planetary design was implemented to reduce backlash. It was used to evaluate systems and operations for the space station, manned maneuvering unit (MMU), shuttle, OMV, and Skylab. Martin Marietta also developed the Viking's surface sampler subsystem described early in this chapter, the Integrated Orbital Servicing System (IOSS), and produced numerous robot studies. All this makes for an impressive resume in robotic studies.

The FTS joints are powered by direct current samarium cobalt motors with harmonic speed reducers (Figures 2.100 and 2.101). Motors for all joints are located in the joints themselves. Wiring for the motors is passed through the rotary joints via twist capsules. A distributed motor, pitch-yaw-roll wrist was chosen and is described in Chapter 3. The Flight Telerobotic Servicer utilizes harmonic drive transmission. Locating the drive motors in the joints produces a very bulky profile: the shoulder's dimensions are 16.5" by 16.5" by 7.5". It also decreased the robot arm's performance, as the joints near the base must carry the weight of the motors beyond it. Even in a weightless environment, such as space, mass is a concern because of inertial effects.

The manipulator kinematics are shown in Figure 2.102. Discrete single-axis actuators are used in a stacked configuration. The impact of this is felt more in the wrist then in the shoulder where greater dexterity is required. Separate fail-safe brakes are incorporated into the design to prevent backdriving in case of power failure. One advantage of discrete actuators is the high torque that is achievable, but at the price of losing dexterity. A disadvantage inherent in this design is cable management. Figure 2.103 shows the rolling cable concept. ABB patented a similar concept in the later 1980s (Gunnarsson 1989).

The dead weight of the yaw and roll actuators is offset by the high load capacity of the individual actuators. The shoulder uses three motors coupled to 100:1 harmonic drive speed reducers (Figure 2.101). FTS designers focused on the two traditional objectives of robot design, namely, precision and load capacity. The more subtle and complex issues of dexterity and reliability are either unaddressed or afterthoughts. Dex-

*Figure 2.99  The Flight Telerobotic Servicer (FTS)*

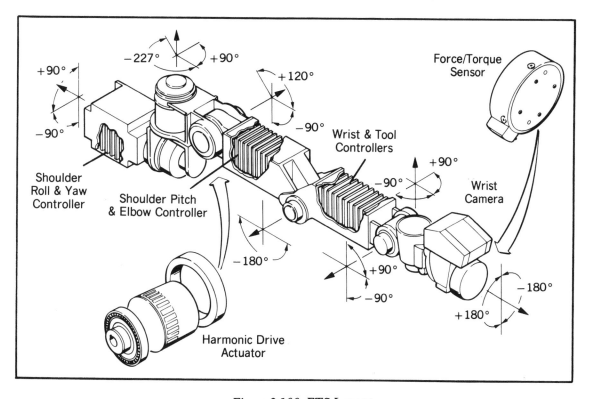

*Figure 2.100  FTS Layout*

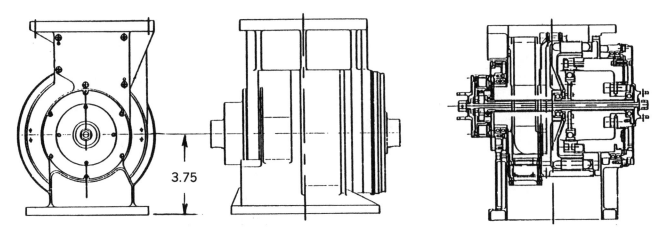

*Figure 2.101  Typical FTS Actuator*

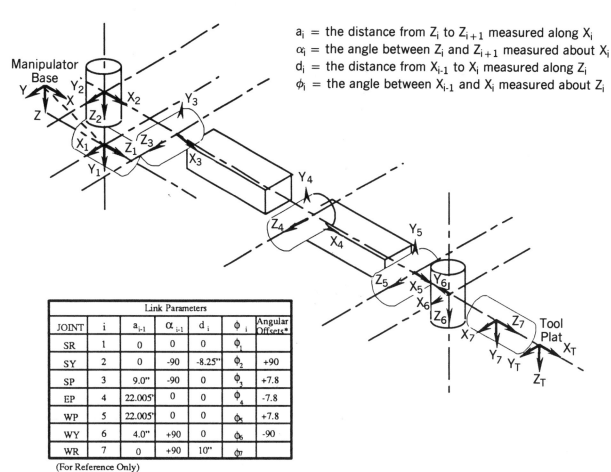

$a_i$ = the distance from $Z_i$ to $Z_{i+1}$ measured along $X_i$
$\alpha_i$ = the angle between $Z_i$ and $Z_{i+1}$ measured about $X_i$
$d_i$ = the distance from $X_{i-1}$ to $X_i$ measured along $Z_i$
$\phi_i$ = the angle between $X_{i-1}$ and $X_i$ measured about $Z_i$

| | | | Link Parameters | | | |
|---|---|---|---|---|---|---|
| JOINT | i | $a_{i-1}$ | $\alpha_{i-1}$ | $d_i$ | $\phi_i$ | Angular Offsets* |
| SR | 1 | 0 | 0 | 0 | $\phi_1$ | |
| SY | 2 | 0 | -90 | -8.25" | $\phi_2$ | +90 |
| SP | 3 | 9.0" | -90 | 0 | $\phi_3$ | +7.8 |
| EP | 4 | 22.005' | 0 | 0 | $\phi_4$ | -7.8 |
| WP | 5 | 22.005' | 0 | 0 | $\phi_5$ | +7.8 |
| WY | 6 | 4.0" | +90 | 0 | $\phi_6$ | -90 |
| WR | 7 | 0 | +90 | 10" | $\phi_7$ | |

(For Reference Only)

\* - angular offsets to account for mechanical alignment in the above figure resulting in zero position sensor measurements.

*Figure 2.102  FTS Kinematics*

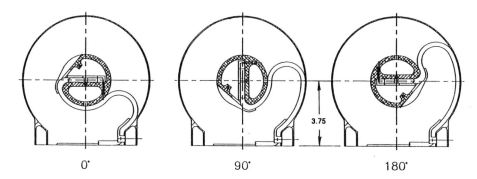

*Figure 2.103 FTS Rolling Cable Management System*

terity is a critical element in unplanned maintenance tasks where a human-like flexibility may be required. The precision requirement is particularly ironic as the arm is controlled by teleoperation. Can an astronaut move his hand in .001" increments? A simple design on the surface, it has several major design flaws that the industrial robot makers have avoided. With an expected working life of 20 years these flaws will become more obvious as time goes by. The distributed actuators create cabling problems in the wrist and shoulder. Harmonic drives are notorious throughout the robotics industry for cracking of the flexspline cup due to shock loads. Perhaps the greatest tragedy of all is the opportunity FTS presented for advancing America's industrial robot technology. Instead a robot was created that reflects with great precision the NASA bureaucratic machine.

**Double AI Arm**

Developed for NASA Langley Research Center and NASA Jet Propulsion Laboratory the Double AI was conceived by Pete Spidelere and Dr. Anton Betzy of NASA JPL and implemented by engineers at Double AI Corporation, Huntvalley, Maryland (Figure 2.104). The design philosophy was to use modular, off the shelf components connected by composite tubes (tube length could vary). Seven degrees-of-freedom and approximately 40 Lbs (18 Kg) of lift was achieved. It resembles the Robotics Research Arm in its use of harmonic drives for the roll-pitch-roll shoulder and elbow, although it is not as well packaged. The four-axes roll-pitch-yaw-roll wrist is intended to decrease singularity albeit at the price of an additional axis. A fiber optic slip-ring was developed for unlimited tool plate roll.

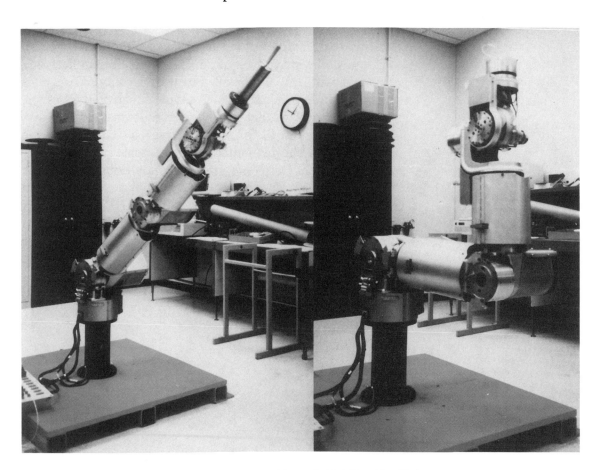

*Figure 2.104  Double AI Arm*

## The Dextrous Arm Design

None of the past or current designs could be called truly anthropomorphic, in either their kinematics or their structures. To achieve anthropomorphism, the kinematics of the robot arm must equal the kinesiology of the human arm. The Dextrous Arm design is the first true synthesis of structure plus kinematics, and has produced the man equivalent robot arm described below. The U.S. patent was filed January 3, 1989 (Rosheim 1990a). What is offered is a general solution to the problem of mechanical robot arms. Although other arm types may be better suited for specific applications, none offers the adaptability to multiple purposes found in this arm. The transfer of my initial wrist technology to shoulder design has made the Dextrous Arm concept possible. Since the shoulder carries the greatest load of all the joints in a robot arm, its potential flexibility is problematic but has been enhanced in the present design by the use of a three degrees-of-freedom, pitch-yaw-roll joint and through placement of all arm motors adjacent to, but not on, the joints themselves. Simplification of the joint through modularity and unique decoupling techniques within the wrist itself are other state of the art features. The use of speed reducers in the shoulder roll, elbow, and wrist increases load capacity, while the two available wrist joint designs provide high precision motion for the endeffector. The Dextrous Arm's shoulder improves on conventional designs by having up to 180 degrees of singularity-free pitch and yaw motion about two axis centers as in the human shoulder (Figure 2.105).

The remote location of the pitch, yaw, and roll motors also improves dynamic performance because of the reduced load carried by the shoulder. The arm is able to move faster and lift greater loads because the motor's weight is distributed toward the rear of the arm (Figure 2.106). The elbow drive is shown in Figure 2.112. Range of motion and dexterity is equivalent to the human arm (Figure 2.107). This is the first time that this has been achieved. Load capacity is 25 Lbs (11.4 Kg).

The 9" diameter shoulder is shown in Figure 2.108. The joint is basically two gimbals coupled together. The individual joints are singularity-free, acting like a ball-and-socket joint. One joint is inadequate to produce a practical amount of pitch/yaw range; therefore, two are geared together, doubling the range of motion. To visualize its construction, imagine a mirror slid between the two joints: the reflection of the joint would be identical to the position of the top joint. It is possibly the highest precision 180 degree range pitch-yaw joint yet designed. Extremely low backlash is achieved by having widely spaced gears on common frames and by using state-of-the-art gear manufacturing techniques. Spur gears offer the only practical means of power transference between double jointed mechanisms. They can be preloaded against backlash with negligible increase in friction, are rugged, and can be manufactured readily. Modular

Figure 2.105  The Dextrous Arm

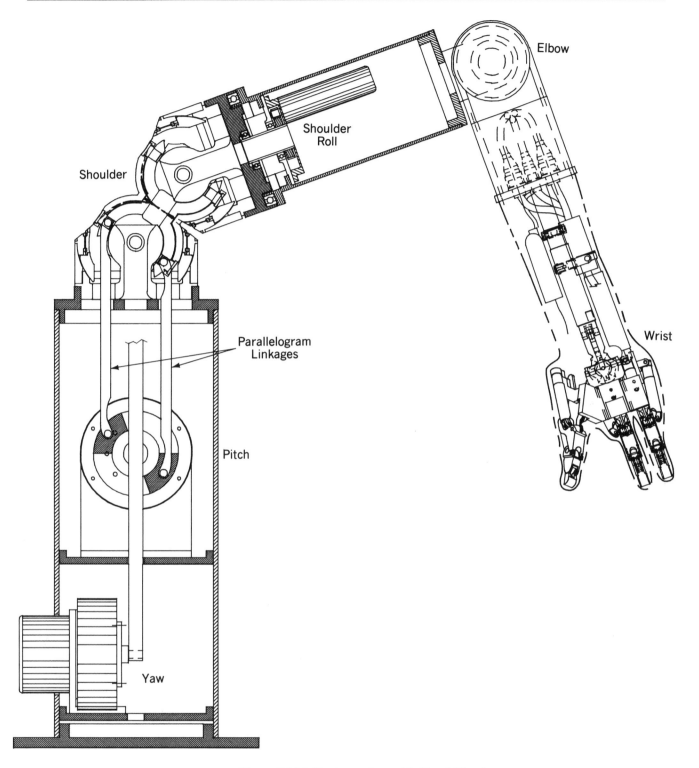

*Figure 2.106 Dextrous Arm with Omni-Hand*

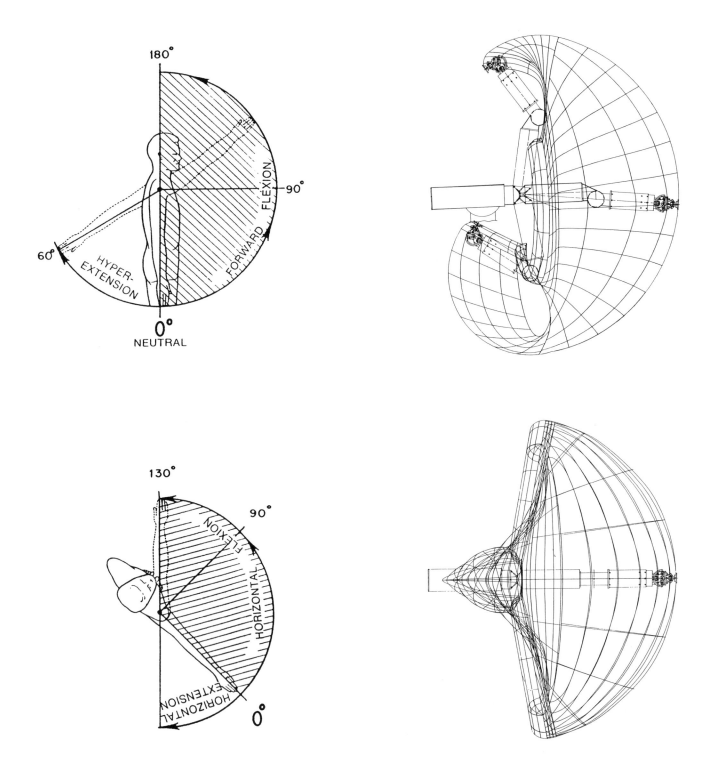

Figure 2.107 Range of Motion Comparison

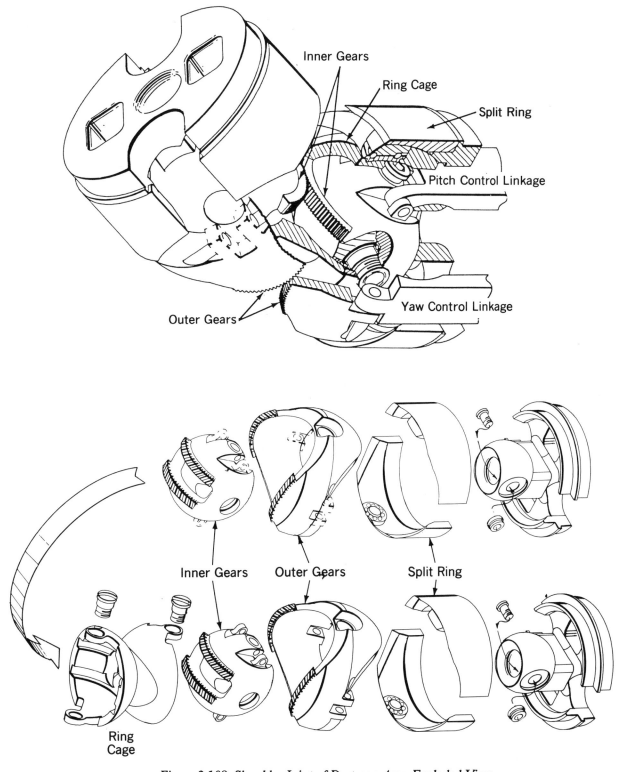

*Figure 2.108  Shoulder Joint of Dextrous Arm, Exploded View*

in design, the shoulder joint may be produced using only a few molds. Because of this joint's high load capacity, it is optimally suited as a shoulder joint.

Two parallelogram linkages power the shoulder (Figure 2.109). A conventional joint would only be able to produce up to 90 degrees of motion from linkages, but the shoulder's unique 2:1 ratio multiples this to 180 degrees in a simple, direct manner. However, a penalty is paid for this dexterity. The joint's output torque is one-half the input torque because of the 2:1 gear ratio. This loss is compensated for by the drivetrains high speed reduction. Directly coupled, cycloidal speed reducers provide drive power. Nonbackdriveable, they provide built-in speed reduction and braking in case of power loss. Self-braking obviates the need for additional brakes, like those used in FTS, and eliminates the need for additional costly maintenance, such as that required for gearboxes or harmonic drives. The pitch control rods are connected to the outer gear, which is supported by bearings on the split ring. Pitch motion is transferred to the upper joint by two pairs of gears which are held on their centers by a ring cage housing for parallel stability. The pivot base attaches inside the ball shell. The yaw push/pull rod connected to the lower inner ball shell produces yaw motion by causing the shell to pivot. The pivoting lower shell pulls the outer gear/housing assembly in an arc perpendicular to itself.

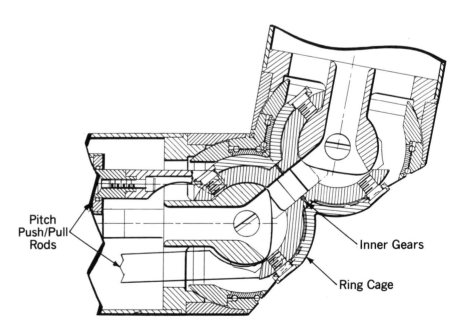

*Figure 2.109 Parallelogram Link Drive (Pitch)*

The shoulder possesses unique kinematics that make it singularity-free. The pitch and yaw axes are decoupled, allowing compound motion to occur. Roll is also decoupled and independent from pitch and yaw. The human shoulder provides the model for this decoupling. The outer gears in the yaw axis form a stable rectangular frame (Figure 2.110). Regardless of their rotation relative to each other, the yaw axis gears maintain their parallel relationship.

In contrast to the FTS shoulder, where pitch and yaw occur in separate planes, the Dextrous Arm's shoulder has pitch and yaw intersecting in the upper and lower joint, which produces a more fluid motion like that of the human shoulder.

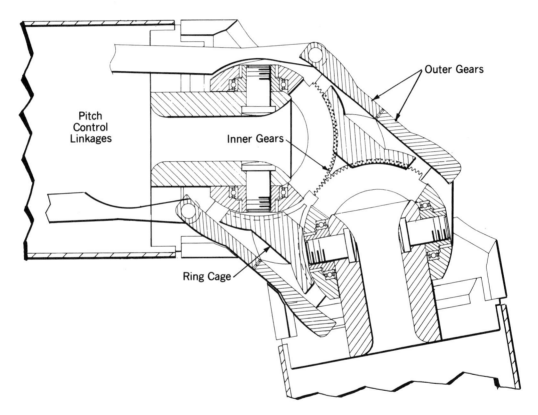

*Figure 2.110 Yaw Axis Outer Gear Frame*

Roll rotation is accomplished by the addition of a roll motor that is driven by offset gearing through a harmonic drive speed reducer (Figure 2.111). Another option is to rotate the entire joint (Figure 2.112). Because the shoulder joint is a high angulation, constant velocity U-joint, roll motion maybe transmitted through it at any pitch/yaw angle. This is valuable in a situation in which the shoulder motors need to be housed in

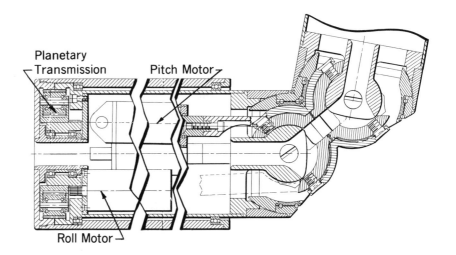

*Figure 2.111 Rotation of the Shoulder for Roll*

the base. A motor located in the trunk drives the bearing supported shoulder tube through a planetary transmission, which provides speed reduction and increases torque. This feature eliminates the need for a roll motor between the shoulder and the elbow, which would put an extra weight burden on the shoulder and increase wiring through the shoulder.

A feature of this joint in its 9" diameter configuration is a 1.5" through hole. This is critical, as it allows smooth passage of the wiring for the elbow and wrist. External wiring would encounter flexing and potential abrasion that could lead to a wiring failure. This is a significant improvement over the FTS which requires special cable handling on each axis.

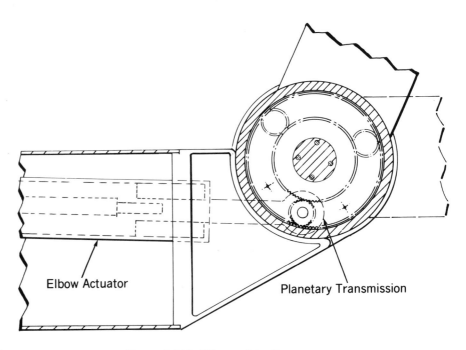

*Figure 2.112 Elbow of the Dextrous Arm*

A sharp contrast is seen when comparing this shoulder joint to others. The IRB 1000 Arm and Tetrabot have half the range of motion in pitch and yaw because they use only one gimbal joint. The Intelledex Arm and LTM are somewhat better with only one axis deficient when compared to the Dextrous Arm's shoulder.

Modularity is a key feature with the arm, as each joint can be easily removed for maintenance or replacement. However, it is not modular in the sense that each joint is interchangeable, as in LTM, because by definition the shoulder, elbow, and wrist are unique mechanisms providing distinct capabilities.

The Dextrous Arm's elbow emulates the human elbow in several ways. The typical robot elbow is little more than a pivot bearing driven by a linear actuator, analogous to the human biceps and forearm muscles. In an effort to increase precision, some designers locate the motor at the elbow, which creates load distribution and sizing problems. In contrast, the Dextrous Arm increases the precision of the elbow without adding the mass of a direct drive motor. A worm gear speed reducer is used for the elbow. This cost effective approach also provides for self-braking in the event of power failure. Range of motion is greater than the human elbow and it may bend backward, unlike its human counterpart. Another approach is an interesting hybrid between a direct drive and a remote elbow, employing elements of both. The joint is driven by a screw/nut linear actuator mounted on a pivot in the upper arm. A planetary gear system translates the linear motion to 110 degrees of arcuate motion, only 40 degrees less than that of the human elbow (Figure 2.114).

This arm is a hybrid and shares the features and limitations of the revolute and serpentine arms that it is descended from. The result is a man equivalent arm that shares the strengths and weaknesses of the human body. Precision and load capacity are not as high as in revolute arms but higher than in serpentine arms. Dexterity is less than the serpentine arm's, but greater then the revolute arm's. Load and precision are less than a single revolute arm, but its greater dexterity provides an overall greater average precision due to its singularity-free design. The shoulder joint doubles the output speed but reduces torque by one-half. This makes it like the human arm; fast but with limited torque. Thus we see the birth of a new breed of arm which has both human and robot traits. The human-like kinematics of the arm require human-like controls to reach its full potential. To fully utilize the mechanical potential of the Dextrous Arm, further advances in motion control and sensors are needed. The University of Colorado at Boulder is developing software to control the arm (Steffen 1993 and 1994). Ross-Hime Designs has constructed an articulated mock-up and completed engineering for a powered unit.

## Conclusion

Examination of robot's arms was begun by tracing the evolution of machine tools and automobiles. The human arm's kinesiology and overall design was used as a guidepost for evaluating styles and trends in industrial arm design. The negative impact singularity has on arm performance was reviewed. Standards for dexterity and precision were discussed as tools for evaluation. Following this, elbows and traditional as well as two new arm morphologies provided us with a baseline for other arms. Design considerations for backdriving, counterbalancing, actuators, and components were examined. While stock power sources and mechanical components were once used, customized actuators and mechanical components unique to robotics are now used. The origins of industrial robots arms in the 1930's and their structural and kinematic progression from simple cylindrical and polar coordinate arms to anthropomorphic designs were revealed. Sidetracks on the path to anthropomorphic designs, such as serpentine and hybrid concepts, were also examined.

Efforts in modularity, with the as yet unfulfilled promise of simplification, was shown in the LTM and FTS. The many faltering attempts made by multimillion dollar designs that are well engineered but poorly designed indicate the weak grasp of structural and kinematic fundamentals in otherwise well trained engineers. New applications and markets are being opened by anthropomorphic technology. These opportunities appear to be as numerous as the uses of the arm itself. The next plateau to be surmounted in the robotic frontier is that of controls, sensors, and software.

# 3

# *Wrists*

The wrist is the first mechanism developed especially for robotics. While it resembles several different mechanisms such as universal joints, gyros, and differential joints, none have all of its unique features. Originally created to meet the need for fine orienting of robot arm endeffectors, new actuators for the shoulder, spine and legs may be derived from a knowledge of wrist structure and kinematics. Almost all of my technology—shoulders, hands, and other joints—was derived in some part from the wrist.

The robot wrist, like the human wrist, is located at the end of the robot forearm, where it manipulates the endeffector and thus the work piece. Wrists fall under the classification of three-dimensional space mechanisms, the most complex discrete mechanisms in existence. New demands in unstructured environments require improved wrist actuators. A service application for improved robotic wrists is cleaning bathrooms in which tool manipulation requires brushing over and around complex three-dimensional shapes.

Without a wrist, a robot arm is little more than a miniature crane. This can be easily demonstrated by extending one's own arm, omitting all wrist motion, and limiting arm motion to up and down, rotating with your trunk while attempting to perform a variety of functions normally requiring wrist movement. The importance of the wrist joint quickly becomes apparent.

The wrist, in fact, is essential to the creation of robot structures capable of high performance, human-like functions. For this reason, a robot wrist with human-like flexibility and precision is the *sine qua non* for a high performance robot.

In a conventional six axes robot arm, three of the six axes are in the wrist. Three degrees-of-freedom wrist joints are essential in teleoperators, industrial robots, and hybrid telerobots that require dexterity, precision, and simplified master-slave interfacing. Robot wrists need these features to operate effectively in the complex three-dimensional environments that humans create and live in. For simpler tasks, wrists with fewer degrees-of-freedom or less range of motion may be derived from three degrees-of-freedom wrists.

The state of the art industrial robot wrist requires three degrees-of-freedom and at least the range of motion of the human wrist in order to provide the dextrous, three-dimensional manipulations required for arc welding, paint spraying and sealant application. Wrist dexterity is defined as the amount of wrist range in the axes and the absence or mitigation of singularity in the workspace.

Presented in this chapter is a review of the human wrist as a guide to the design of robot wrist actuators. The first patented robot wrist is also examined and two well established designs that have seen countless hours of service are critiqued. For a more comprehensive treatment of this subject refer to *Robot Wrist Actuators* (Rosheim 1989b).

### The Human Wrist

Using the example of the human wrist, well designed robot wrists have their actuators located in the upper portion of the arm or in the base of the robot. This produces a robot wrist that is compact and lightweight. The wrist, also known as the radiocarpal joint, provides a flexible interface between the forearm and hand (Figure 3.1). It is composed of eight bones, the midcarpal joints, that are held together by a complex network of tendons. This ball shaped mass of bones rides on the socket shaped cartilage of the forearm's radius and ulna bones. The wrist is driven by antagonistic pairs of tendons powered by muscles in the forearm. Wrist roll is provided by muscles in the forearm that crisscross the radius and ulna and, in anatomical terms, provide supination/pronation. Additional roll, medial and lateral rotation, comes from muscles in the shoulder area that rotate the humerus. The human wrist is driven by muscles located away from the wrist, and by tendons and bones in the arm. Together they supply the power to drive the wrist's compact group of bones. This feature allows for larger, more powerful muscles than could be located in the wrist itself.

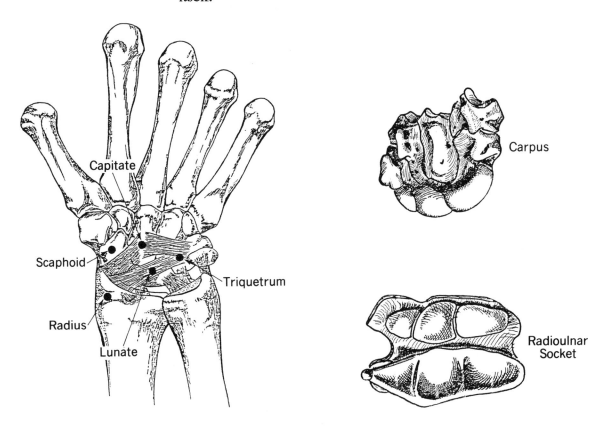

*Figure 3.1 Bone Structure of the Human Wrist*

The human wrist (Figure 3.2) is capable of moving the hand 70 degrees in the radial/ulnar deviation mode and 170 degrees in the flexion/extension mode. The operational modes correspond to pitch and yaw motion that is singularity-free. Figure 3.3 illustrates the kinematics of the human wrist.

When the wrist moves more than 45 degrees off center its ability to transmit roll degenerates (gimbal lock) as in the case of extreme flexion/extension. The hand is rotated, but in a sweeping motion, scribing an arc about the roll axis. Maximum wrist rotation by the arm is 270 degrees total. This comes from two sources: the radius and ulna in the forearm crisscrossing each other (supination/pronation) and rotation of the humerus by the shoulder joint (medial and lateral rotation).

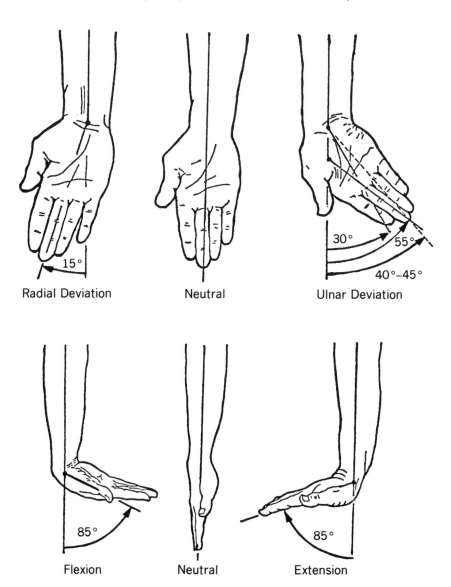

*Figure 3.2 Kinesiology of the Human Wrist*

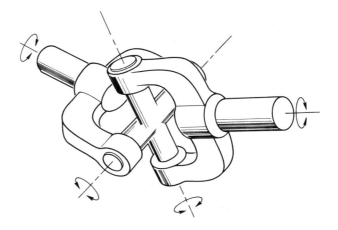

*Figure 3.3 Kinematics of the Human Wrist*

## Robot Wrist Overview

Wrist actuators were first invented for spray painting robots in the 1930s, and in the mid-1940s teleoperators were developed at Argonne National Laboratory for handling hazardous radioactive materials. Since the advent of industrial robots in the 1960s, more sophisticated designs with greater dexterity have been developed. Robot spray painting and welding place new demands on wrist dexterity. With the demise of hydraulic robots, virtually all wrist actuators are powered either directly or remotely by electric motors.

Robot wrist actuators typically are divided into two categories, depending on the orientation of their axes: pitch-yaw-roll wrists like the human wrist; and roll-pitch-roll wrists. Using the terminology of the shoulder, the three degrees-of-freedom are pitch, yaw, and roll. These robotic terms correspond to the human wrist's kinesiologic terms of flexion/extension, radial ulnar deviation, and supination/pronation (Figure 3.4). The pitch and yaw axes are at right angles to each other and provide pivotal movement for the wrist, while their range generally is 180 degrees (an industrial standard). Roll rotates the end effector and can vary from part of a revolution to 360 degrees of continuous rotation. A tool plate on the front end of the wrist supplies a means for attaching end effectors to the wrist. Pitch-yaw-roll wrists are singularity-free (see definition below) and are more flexible than other types, but they are also more complex. They require continued efforts to simplify them and thereby reduce cost while increasing reliability.

Roll-pitch-roll wrists (Figure 3.5) are much more numerous than pitch-yaw-roll wrists in robotics and teleoperator systems. In many cases their greater simplicity provides a more rugged design. However, they have a serious mechanical problem - singularity.

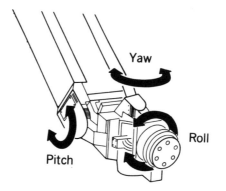

*Figure 3.4  The Three Degrees of Freedom*

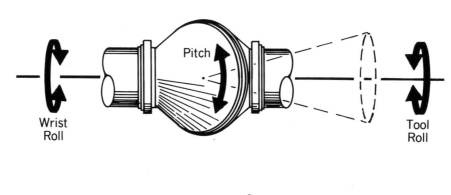

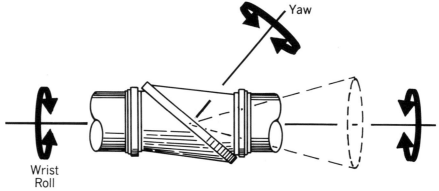

*Figure 3.5  Roll-Pitch-Roll Wrists*

## Wrist Singularities

As in shoulder, singularity in the wrist is the area in the workspace that has to be avoided because of mechanical design defects. Singularity causes the joint to jam as it attempts to move into an area in which it has no axis of rotation or range of motion. A dynamic problem, singularity becomes most apparent when the wrist is in operation. In many applications, singularity may cause very complex programming and tooling problems and has caused damage to equipment and tooling. In addition, unpredictable, high velocity "joint flipping" can lead to operator injury (Paul and Stevenson 1983, Milenkovic 1984, 1987a, Trevelyan 1986, McKinney 1988). Most robot joints that have singularities are the roll-pitch-roll type. Robotic wrist design is dominated by two common roll-pitch-roll joints that for all practical purposes are kinematically identical: the split tube, and the still older clevis type joints. Split tube designs have been traced as far back as 1918 (Wendler 1918, Minium 1918). A popular variation of the split tube design couples multiple split tubes in series (Figures 3.6 and 3.6A). While this increases pitch, it does not decrease the singularity (Dahlquist 1987, Dahlquist and Kaufmann 1987, Tsuge 1991, Zimmer 1987a,b,c, 1988).

However, neither of the above type is able to move the endeffector in an axis perpendicular to the pitch axis (yaw) without "flipping" the wrist's roll axis. Since this cannot be done instantaneously, a conical void in the workspace is generated. For example, when the wrist moves at full extension from the pitch attitude (Figure 3.7) to the yaw attitude the entire wrist must be rotated up to 90 degrees about the forearm axis (Figure 3.7A). The inefficiency of this corrective movement can be visualized by moving your hand vertically in flexion/extension (pitch) and then using supination/pronation (roll) to reposition the wrist into a quasi radial/ulnar deviation (yaw). An example of a mundane activity made awkward by such a limitation is painting a wall with vertical strokes utilizing pitch motion. Any small shifts by the wrist to the left or right (yaw motion) will cause jamming or slow the application while the wrist is flipping. The catch phrase "it's all in the wrist" is literally true. For a robot, this dwell time means the wrist will not move smoothly. A critical conical area in the wrist's work volume has to be avoided, adversely affecting wrist dexterity. Located at the center line of the forearm, the singularity interferes with operation of the endeffector, whether it is a welding tongs, a sealant applicator, or a paint gun. Two examples of pitch-yaw-roll wrist designs that have singularities in their workspace centers are by Tesar and Akeel (Tesar 1989, Akeel 1989).The still popular and economic walk through programming method is also difficult with this type of wrist because of incompatibility with the human operator's own pitch-yaw-roll wrist. According to ABB GRI engi-

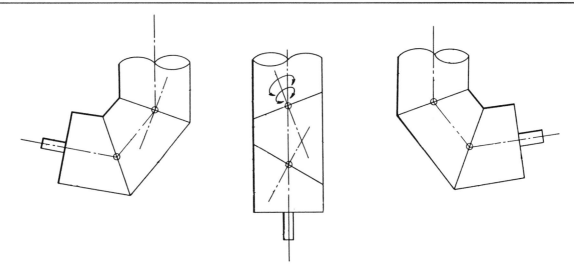

*Figure 3.6  Multiple Split Tube Wrist*

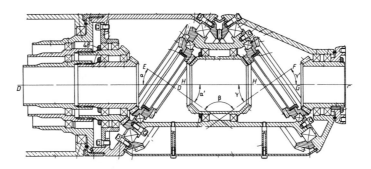

*Figure 3.6A  Multiple Split Tube Wrist, Sectional View*

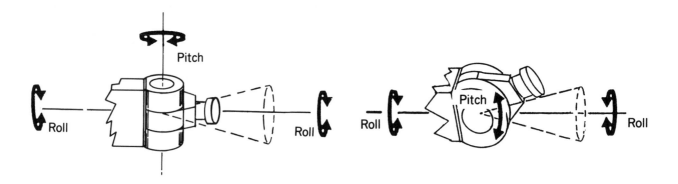

*Figure 3.7  Roll-Pitch-Roll Wrist in Pitch Mode*          *Figure 3.7A  Roll-Pitch-Roll Wrist in Yaw Mode*

neers, walk through programming can cut robot programming time in half over conventional teach box or offline programming.

In contrast, properly designed pitch-yaw-roll joints are free of singularity in their operating range. Based on universal joint principles in the human wrist, they are able to position the endeffector without flipping either the pitch or yaw axis because of their two true perpendicular axes. This advantage makes them more dextrous, allows backdriving for teleoperation and walk through programming, and eliminates the safety and tooling problems generated by singularity.

One of the oldest pitch-yaw-roll wrist configurations is stacking single axis actuators on top of each other. However, this design is inflexible because of the different independent radii of rotation (Figure 3.8). Toggling occurs as the wrist moves in circumduction (a sequence of pitch and yaw motion). Observe the toolplate's wildly uneven path.

Another wrist type based on single U-joints has inadequate range of motion (Figure 3.9). Two types of wrist designs use U-joints. The first consists of a modified universal joint with portions of the upper fork cut away. The second uses revolute joints that are either directly or remotely driven through bevel gears. Offset mounting of the actuators reduces the overall length and allows continuous yaw rotation, but introduces singularities in the wrist operation when the pitch and yaw axes intersect.

These two wrist types are limited to an effective range of 180 degrees of pitch and 90 degrees of yaw because of the singularity condition known commonly as "gimbal lock." The degree-of-freedom loss occurs because of the single intersection point of the pitch and yaw axes and interference with wrist members. Most evident when the wrist is "hardover," or operating near its limits, the pitch or yaw axis locks up or hits the support structure carrying the upper members. Wiring or "cable management" of the pitch and roll motors for both power and position/velocity feedback, in addition to any additional sensor or power equipment for endeffectors, becomes a serious problem. The wiring must pass though several small radii that subject it to flexing whenever the wrist is in operation.

Another problem is that roll transmission through the wrist is impractical. Roll motion is usually supplied by an additional motor at the toolplate that requires the passing of wires through the wrist. Many designers have tried to increase the range of motion by removing an arm from the upper or lower clevis, but this does not change the basic kinematics nor does it allow the wrist to achieve 180 degrees pitch and yaw. For a simple example, picture a ball-and-socket which has one common center for pitch, yaw and roll (Figure 3.10). Can it have 180 degrees of pitch and yaw? No, because the ball's support shaft will interfere with the socket's lip (analogous to the structural elements of the universal joint) that holds the ball in place. This example also applies to single universal joint wrists (Figure 3.11). The upper clevis collides with the lower clevis during circumduction. This

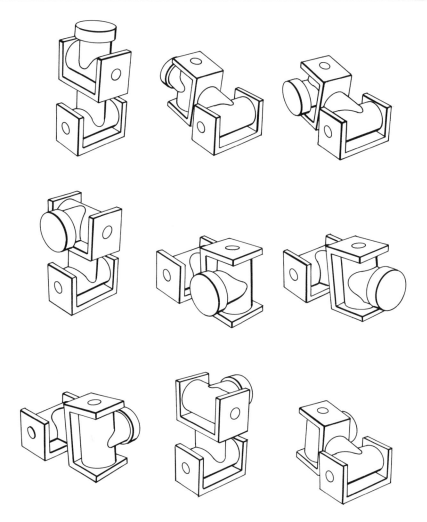

*Figure 3.8  Pitch-Yaw-Roll Wrist Moving in Circumduction*

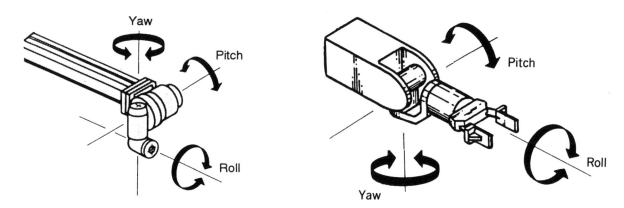

*Figure 3.9  Single Centerpoint Wrists*

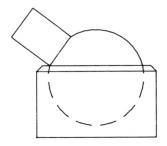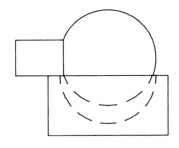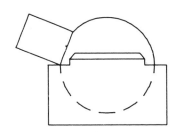

*Figure 3.10  Ball-and-Socket Test*

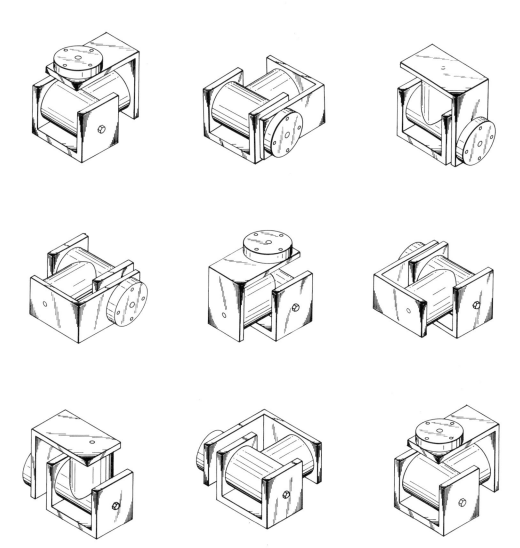

*Figure 3.11  Limitations of Single U-Joint Wrist*

structural limitation, as well as the difficulty of driving such an arrangement, illustrates the problems with single center point wrists.

Next to be examined are specific wrist examples. A mixture of chronology and an increasing level of sophistication will be used to trace the evolution.

### Position Controlling Apparatus Wrist

The first robot wrist patent (Figure 3.12), "Position Controlling Apparatus" by Willard Pollard, was filed April 22, 1938 and issued June 16, 1942 (Pollard 1942). Designed for spray finishing, it was coupled to a four axes robot arm (see Chapter 2 for further information). The inventor took his cue from nature and obviously based his design on the human wrist. This singularity-free ball-and-socket wrist was in some ways ahead of its time. Pitch and yaw is driven by linear actuators that connect to the ball-and-socket end of tie rods. Roll was transmitted through the center of the main ball via a flexible coupling. Flexible driveshafts were also used to drive the pitch and yaw linear actuators. Flexibility was obviously the

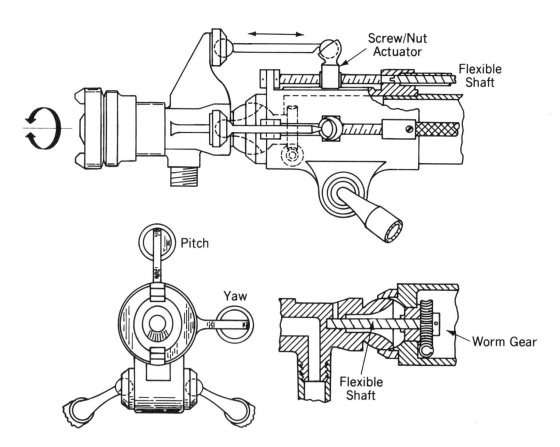

*Figure 3.12  Wrist from "Position Controlling Apparatus"*

primary consideration here and it would appear that the human wrist at least in part was the model for this design. The ball-and-socket simulated the human wrist's radius and ulna cup and the midcarpal joint's hemispherical formation. Tendon-like tie rods powered pitch and yaw while the linear screw/nut actuators simulated the forearm muscles. Like the human wrist's radius and ulna, the wrist's tie rods crisscross during roll motion. A better solution would have been to decouple this motion through bearings that allowed constant rotation.

## Unimate Wrist

The Unimate Wrist was one of the first industrial robot wrists put into mass production. Introduced in 1962, it has seen millions of hours of operation and is the grandfather to many of today's wrists (Figures 3.13 thru 3.15). The 2000B wrist, originally configured like the Planet Corporation wrist for pitch-roll, provided an additional roll axis option which made it into a pitch-yaw-roll wrist. Load capacity of the wrist is 300 Lbs (136.4 Kg) and contributes to an overall robot repeatability of .05" (.002 mm) (SME 1983). A passage through the wrist provided integral air pressure for endeffectors. Two rotary shafts terminate into bevel gears that drive the wrist's pitch and roll. The roll gear is a double gear which meshes with the tool plate roll shaft gear. For the additional axis, an auxiliary drive shaft is added outside the arm which provides rotary power through a universal joint to a point on the pitch axis. Another shaft drives a pair of bevel gears through the additional axis gear box. Setting the backlash is a complicated procedure with a large number of steps. Ultimately, the backlash in the gears and other connections was set by gravity, using the wrist's weight to preload the gears. The Unimate was not a lightweight, the wrist alone weighed 45 Lbs.

The additional roll axis option made the first roll axis into a yaw axis. This joint is powered by a separate hydraulic motor on the left side of the boom via gears, chains, and sprockets. An auxiliary outboard drive shaft transfers this rotary motion through universal joints intersecting with the axis of the pitch joint. Standard torque is 1000 in-Lbs (pitch), 600 in-Lbs (yaw), and 800 in-Lbs for roll. Load capacity is 300 Lbs. The millions of hours these wrists have endured gives proof of the design's reliability and has spawned countless imitations.

In spite of bevel gears, numerous mechanical connections, and a poorly integrated roll axis, the Unimate wrist is a good example of a complicated design that has proven itself useful beyond a doubt. The drivetrain in the arm is also quite complex. It converts the hydraulic cylinders' linear motion to rotary motion through spring loaded rack and pinions, chains, shafts and gears, and also splines. A double U-joint system supplies power regardless of elbow extension or retraction.

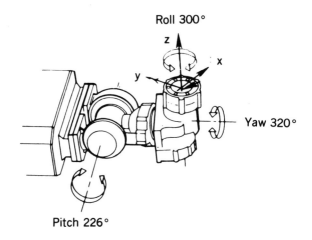

Roll 300°

Yaw 320°

Pitch 226°

*Figure 3.13  Unimate 2000B Wrist*

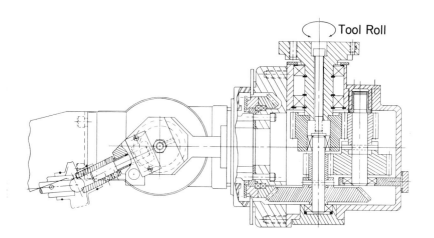

Tool Roll

*Figure 3.14  Unimate 2000B Wrist Detail Roll Axis*

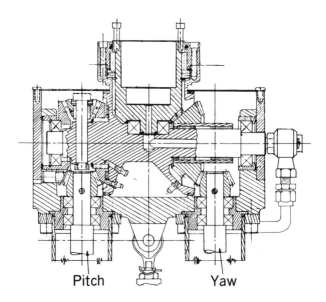

Pitch          Yaw

*Figure 3.15 Unimate 2000B Sectional View without Roll Axis*

## PUMA Wrist

Borrowing the basic design of the Unimate wrist, the PUMA 562 roll-pitch-roll design is well known through countless laboratory and industrial installations (Figure 3.16). Within the wrist itself, bevel gears and pinions, as in the Unimate wrist, are used to transfer pitch and tool roll motion. The pitch shaft terminates in a single stage gear reduction that drives the bevel gear set. Tool roll is driven by a similar bevel gear set that drives a second pair of bevel gears for another speed reduction to the tool plate output shaft. Wrist roll is 120 in-Lbs, pitch 110 in-Lbs, and tool plate roll 120 in-Lbs. Repeatability is .004" (.15 mm). Three motors at the top of the forearm drive the wrist through a Heli-Cal Flexible coupling

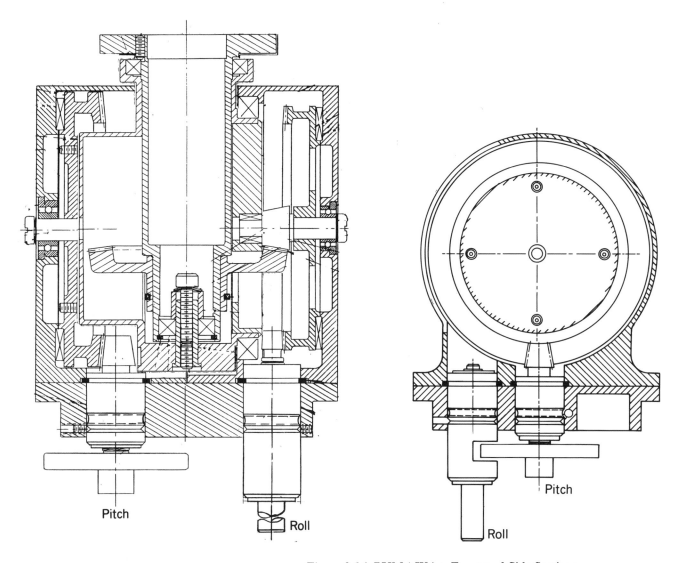

*Figure 3.16  PUMA Wrist: Front and Side Sections*

(Figure 3.17). The flexible couplings are needed to compensate for shaft misalignment and interference constraints caused by the forearm's tapered structure. The flexible couplings also permit limited wrist roll by crossing each other, as the radius and ulna do in the human forearm. Closely imitating the human wrist's roll design seems to be an unfortunate choice as it precludes continuous rotation, limiting it to 280 degrees. Wrist roll is driven by the left shaft driving a spur gear to a larger sun gear. Wrist roll is driven by the bottom shaft through a two stage gearbox, the final stage ending in the large gear. The inner sleeve rotates on two bearings and the wrist is bolted into this sleeve. Pitch and tool roll pass through the wrist roll gear box so as not to obstruct wrist roll. The large number of gears makes this an expensive and potentially high backlash joint as the 12 gears add backlash with every mesh. Bevel gears are especially difficult to "set" and are the highest maintenance item in the wrist. The helical flexible coupling adds compliance to the system while decreasing power output by adding additional friction to the shaft bearings through increased loading. The status quo somehow survives, although much simpler and less expensive roll-pitch-roll versions have been developed (Rosheim 1989b).

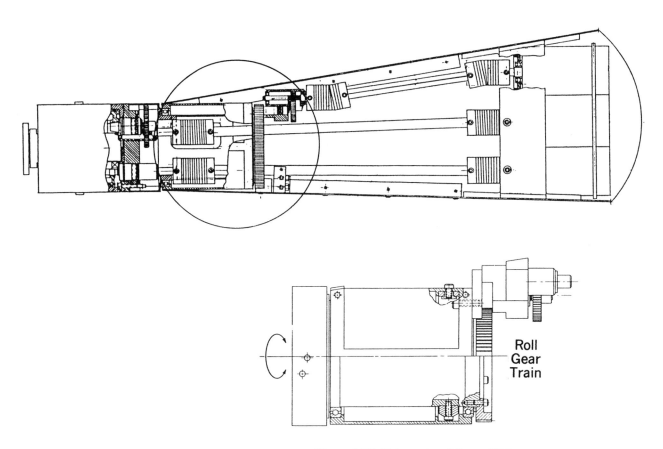

Roll
Gear
Train

*Figure 3.17 Drivetrain of Puma Wrist*

# Direct Drive Wrists

Creating a direct drive wrist simply by mounting motors at the end of the arm has always been an attractive option. Advantages to direct drive include omitting the drivetrain, decreased backlash, reduced weight, and lower cost. Negative side effects include bulky profiles, wiring problems, limited range of motion and lessened flexibility due to design singularities; the latter caused by mechanical interference with the motors and kinematics. This has precluded universal adaptation of direct drive. Presented below are attempts at overcoming the difficulties of direct drive in wrists.

## Dextrous Spherical Robot Wrist

Invented by Oprea Duta and Michael Stansic, the patent was filed May 27, 1988 (Duta 1989). High angulation and singularity-free kinematics are the goals of this design. One common intersection point for pitch and yaw simplifies control by eliminating the offset found in double universal joint wrists.  Development continues  at Notre Dame University.

Pitch is achieved by two semicircular bails pivoting on a horizontal axis. Yaw is obtained from sliding blocks held by the two bails (Figure 3.18). Several methods are proposed for driving the bails, including tendons and up to eight motors, but the most practical is side mounted motors driving the pitch and yaw pivots. Two identical multistage geartrains transfer motor torque to the upper and lower bails while maintaining a phased relationship between the hemispheres.

This design illustrates the difficulties created by pursuing ideal kinematics. The primary drivetrain difficulty is that the upper and lower bails require multistage gearing. Interference of the motors is a problem, although several solutions are offered, they all compromise the design's basic ruggedness. Structurally the design is compromised by the sliding blocks, an expensive method of achieving revolute motion.

## Interlocking Body Connective Joints

Another interesting example of a direct drive joint is the "Interlocking Body Connective Joints," invented by Vladimir Shpigel, and assigned to the Amrus Corporation (Figure 3.19). The patent was filed March 13, 1990 (Shpigel 1992). Two interlocking rings are assembled perpendicular to each other. Each ring's rotates relative to the other ring. Internal passages provide routing of cabling or hoses. Side mounted motors drive internal ring gears that provide valuable speed reduction while increasing torque.

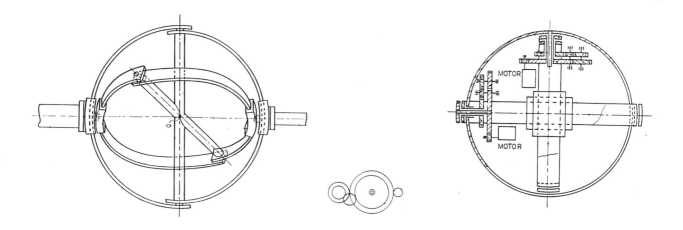

*Figure 3.18 Dextrous Spherical Robot Wrist*

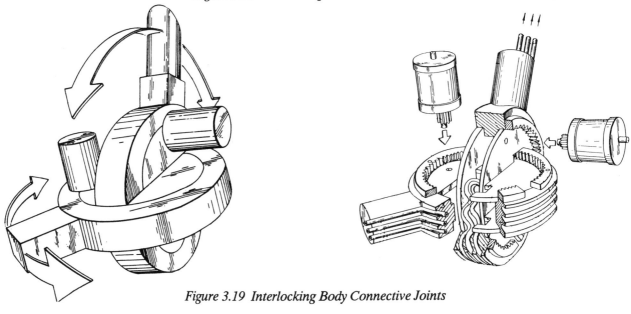

*Figure 3.19 Interlocking Body Connective Joints*

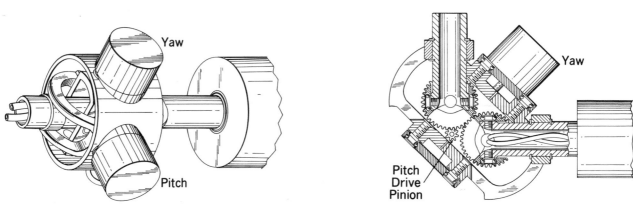

*Figure 3.20 Direct Drive Omni-Wrist*

Kinematically the joint is a single universal joint with the pitch and yaw axis in separate planes. The offset is necessary for the rings to rotate independently. Each axis may move over 180 degrees, however compound rotation is limited. Circumduction motion is limited to $180°$ in one axis and $90°$ in the other axis, because the joint locks beyond this range. Other embodiments of the design include driving the rings through bevel gears that permit mounting the motors on the input and output shafts. Additional gearing methods as well as driving the joint with up to eight motors are described.

## Direct Drive Omni-Wrist

Invented by the author January 1, 1991, simple modification to the $180°$ pitch/yaw Omni-Wrist design results in a direct drive unit. The drivetrain is removed and two motors are mounted perpendicular to each other with the motors' drive pinions meshing with the gear sectors (Figure 3.20). This mesh provides a valuable speed reduction. Wiring, typically a problem in direct drive designs, is a simplified by routing the wiring though the inner gear wrist pivots and out the hollow support shaft. The design already possesses a through hole for wiring or other cabling to pass through the wrist. Applications include: gantry robots, antenna and sensor mounts.

## Flight Telerobotic Servicer Wrists

The Flight Telerobotic Servicer (FTS) Wrists were developed by Martin Marietta and NASA Goddard Space Flight Center for the FTS Arm and were intended to be used during space station assembly (Figures 3.21 and 3.22). They represent a marked regression of wrist kinematic design and compare to concepts 20 years old. The first version uses the classic distributed motor arrangement in a pitch-yaw-roll pattern. Although the wrist is singularity-free, the separate radii of the pitch and yaw axes make for a inflexible design that toggles in circumduction motion (Figure 3.8). The roll actuator is mounted on the output of the pitch/yaw package. Harmonic drives mated to high torque, rare earth motors provide a very high 24 ft-Lbs of torque. This type of wrist configuration was driven in part by the arbitrary performance specifications. Incremental motion precision of .001" drove a design that required encoders mounted directly on the pitch, yaw, and roll output shafts. Also, the 24 ft-Lbs of torque specification made harmonic drives a must. These high torques caused safety concerns resulting in limiting the drives to 7 ft-Lbs. Originally NASA planners must have thought that these requirements would force an

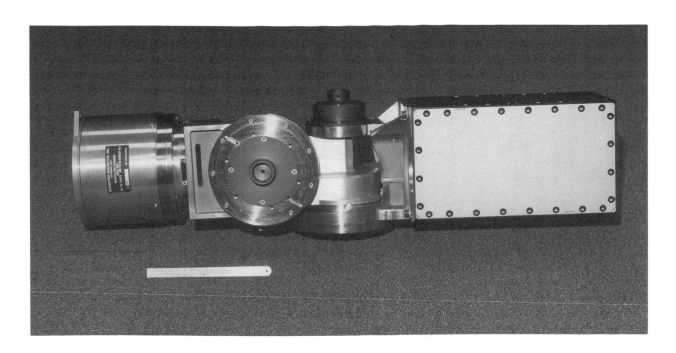

*Figure 3.21  FTS Wrist*

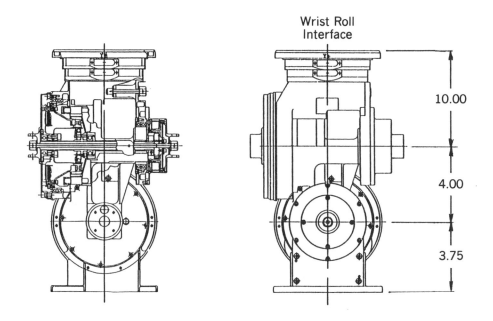

*Figure 3.22  FTS Wrist Without Roll Axis*

advance in wrist technology. Unfortunately, they did not specify dexterity, which was probably poorly understood at the time. The high torques and precision requirements drove the designers back to old technology in an attempt to meet specifications. Ironically, if the wrist singularities are taken into account, the wrist has very poor precision, far less then the original specifications called for.

The following design proposed by Pete Spidelere of Goddard Space flight Center (Figure 3.23) reconfigured the actuators into an offset pitch, yaw, roll design. The yaw actuator is side mounted and provides the benefit of greater range of motion in that axis. Unfortunately, having one common intersection point for the pitch and yaw axes reduces the effective singularity-free motion to 180 degrees pitch and 90 degrees yaw. A problem common to both designs is the enormous number of cables and cable connections (now estimated at 20,000) that must pass through the wrist's pitch and yaw joints. These cables are permanently woven through the wrist and are not *in situ* replaceable items. Like the shoulder, the cables roll in a circular cavity but in a more three-dimensional manner. While the design life is specified at 20 years, one wonders how this can be accomplished.

An additional congressionally mandated goal was to have the FTS project create spinoffs of benefit to the commercial robotics sector. Clearly, these two wrist designs fail in this regard. In both cases the design had been tried and abandoned years earlier by the robotics industry because of reliability problems and lack of dexterity. In 1991 the FTS project was scaled down.

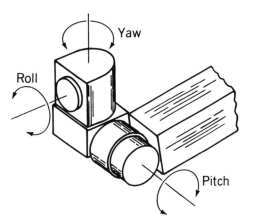

*Figure 3.23 Proposed FTS Wrist*

## The Omni-Wrist

Evolution of the Omni-Wrist design covers a span of over 15 years of research, inventing and interaction with industry which continues to this day. My first design effort resulted in the Hydraulic Servo Mechanism (Rosheim 1980, 1981), a hydraulic ball-and-socket pitch/yaw wrist actuator (Figure 3.24). The goal was to make a wrist with human-like kinematics using as few moving parts as possible. A design resulted that used the flexibility that hydraulics provided. Hydraulic power largely eliminated the need for a drivetrain to transmit power from motors. Power would come from the joint itself, in the form of pressurized fluid acting on vanes, encumbered only by the hydraulic power lines or manifold. This seductive feature continues to attract designers who want high power, simplicity and flexibility. Another factor was that hydraulic robots were prevalent at that time.

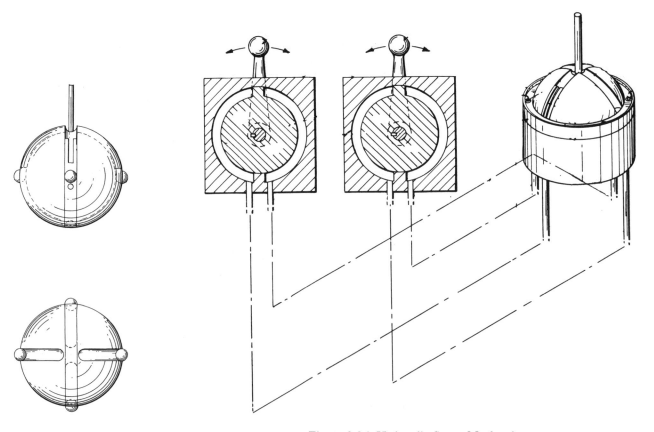

*Figure 3.24  Hydraulic Servo Mechanism*

Basically a ball-and-socket, the ball has two perpendicular grooves machined into it which terminate at the "south pole." In the equatorial plane of the ball, four equally spaced pins inject hydraulic fluid into the ball grooves. The grooves seal against the socket, forming two curved, double acting hydraulic cylinders. The blowout load is counteracted by the weight of the arm, when used as a shoulder; or the workpiece, when used as a wrist.

With the shift of industrial robots away from hydraulics and towards electrics, establishment of range of motion standards, and remote location of motors, I moved towards a remotely driven mechanical wrist (Rosheim 1988b). Robot Wrist Actuator was my first effort at using electrical drives (Figure 3.25). The goal in this design was to make a wrist that could compete with existing mechanical wrists in the areas of cost, simplicity and performance. Design features included a singularity-free pitch-yaw-roll configuration with double pitch-yaw intersection points. The double jointed design was a key breakthrough: it uses the minimum number of joints required to achieve 180 degrees of motion. Adding more joints produces unacceptable backlash, inertia, increased cost and decreased load capacity while introducing nonlinearity. Drivetrain concepts were developed to relocate motors away from the wrist and onto the arm. This design eventually became the Omni-Wrist II, the shoulder joint reviewed in Chapter 2 (Rosheim 1988d, 1990a, 1990b).

*Figure 3.25  Robot Wrist Actuator*

In an effort to further simplify, reduce weight, increase ruggedness, and lower cost I developed the Compact Robot Wrist. The Compact Robot Wrist replaces the circular elements and large bearings with semicircular elements and cam followers found in the Robot Wrist Actuator (Rosheim 1986, 1988c, 1989c). Overall length is reduced by one-half. Kinematically identical to its predecessor, the goal was to make the drivetrain compatible with existing robot arms.

The Omni-Wrist is the commercialized version of the Compact Wrist Actuator (Rosheim 1988). Refinements included a bevel gear drivetrain guided by antibacklash cam followers in the gear bails. Bevel gears allow the mechanical backdriving necessary for simple walk through programming. Figure 3.26 shows the Graco Robotics, Inc. (GRI) version of the Omni-Wrist (Rosheim l991). GRI was purchased by ABB in 1991. Featured on the EDX-1 robot arm, the Omni-Wrist is used for spray finishing, adhesive, and sealant application (Figure 3.26).

The wrist has several features not found in other designs. It overcame the "weak wrist syndrome" of a low power to weight ratio and fragile structure found in earlier designs by offering a 1000 in-Lbs pitch-yaw load and 500 in-Lbs roll load capacity in a wrist diameter of 5.5". Its uniquely compact, singularity-free pitch/yaw motion with continuous bidirectional roll rotation is particularly useful when low dexterity grippers have limited manipulative ability. The 180 degrees of singularity-free pitch and yaw

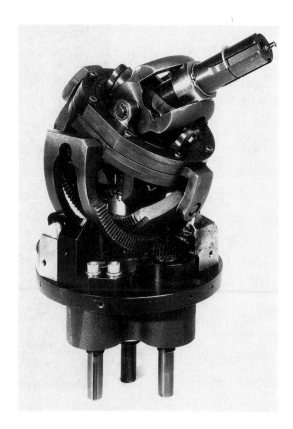

*Figure 3.26  Omni-Wrist used by ABB GRI*

motion is a product of the Omni-Wrist's two separate pitch/yaw intersection points, making it double jointed (Williams 1990, 1992). The double jointing avoids the gimbal lock found in single axis intersection units. Because of its universal motion the pitch, yaw, and roll structure does not package as simply as the PUMA Wrist and requires a flexible rubber boot to protect it in harsh environments. All motors, including the roll motor, are mounted away from the wrist in the rear of the forearm (Figure 3.27). Centrally locating the motors in the forearm eliminates flexible wiring in the wrist, simplifies maintenance and produces a nonbackdriveable, self-braking feature. Roll rotation of the wrist is provided by the universal joint design, which allows the roll motor to be centrally located with the yaw and pitch motor. This creates a more compact design than designs that locate the roll motor at the end of the wrist. The wrist has the valuable design advantage of a hollow 5/8" passage extending through its length.

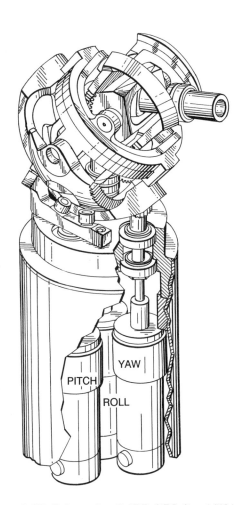

*Figure 3.27  Drivetrain of ABB GRI Omni-Wrist*

The passage allows internal routing of optical fibers, wires, hoses or tendons for human-like hands. In contrast to that of the PUMA, the Omni-Wrist drivetrain and head have fewer parts, gear meshes, and mechanical connections. An additional roll axis shaft, concentric with the first, may be added for powering grippers or tools. Two universal joints on the shaft allow flexing of the wrist.

The observation that these bevel gears in the wrist act like springs when under load led to the development of a NASA Langley contract for stiffening wedges that utilized ball bearings to reduce friction. A customized gearbox in the base provides three stage speed reduction for each axis that produces optimal torque with a minimum of friction (Figure 3.28). Conical rollers were used as cam followers to eliminate tracking error (Rosheim 1991).

*Figure 3.28  NASA Langley Wrist with Stiffening Wedges*

In the latest generation, the drivetrain uses single rods driven by screw/nut actuators. This simplifies the design, lowers the cost, when compared to bevel gears, and creates self-braking in case of power loss. Torque is 1000 in-Lbs for pitch and yaw, and 500 in-Lbs for roll. Support for the bails comes from a one piece, grooved cup derived from the more elaborate four wedges used in the NASA Langley Research Center design (Rosheim 1993a). Dicronite dry film lubricates the ball bearings in the bail grooves and guide cup. NASA Jet Propulsion Laboratory has applied the Omni-Wrist to its mobile hazardous waste robot (Figure 3.29) because it adds the additional dexterity necessary to deal with unstructured tasks and environments.

## Wrist Dexterity

Table 3.1 provides a dexterity rating for roll-pitch-roll wrists, starting with the lowest dexterity to the highest. The final design represents a theoretical ideal. Table 3.2 rates the relative dexterity of pitch-yaw-roll wrists in existence. As with Table 3.1, the final wrist represents theoretical ideal. Additional considerations important to wrist selection include precision, stiffness, and load capacity. As dexterity increases, precision and stiffness generally decrease, but stiffness and precision may vary from one wrist design to another within the same dexterity category.

Pound for pound, the load capacity of direct drive wrists, such as the FTS design, is higher and the precision of each axis is greater, but the overall dexterity is lower. The wrist becomes in the higher dexterity versions a translatory device that coordinates the motors' rotary inputs into an arcuate output.

Design goals for the future include further development of singularity free, high angulation, constant velocity, universal joint type wrists. Real-time tasks in unstructured environments are the largest application base and require human-like flexibility. This is the inverse of the factory environment where everything is structured. Prosthetics, too, would benefit from spinoffs of this technology. The development of a hardshell housing is one area of value. A hardshell is ideal for ease of maintenance and overall ruggedness as shown in Modular Robotic Joint presented in Chapter 2. A larger through hole for a given wrist diameter is sought to accommodate a greater number or diameter of cables, fiber optics, and hoses. Increased mechanical efficiency and a more advantageous gear ratio such as 1:1 over the 2:1 ratio found in the Omni-Wrist would be desirable. This may require abandonment of the double U-joint design core radical concepts such as shape memory alloy or even metal alloys as depicted in the science fiction movie Terminator 2.

*Figure 3.29  Latest Generation of Omni-Wrist*

*Table 3.1  Relative Dexterity of Roll-Pitch-Roll Wrists*

| | | | |
|---|---|---|---|
| | | | |
| *Clevis Single Center* | *Split Tube* | *Multi-Split Tube* | *Multi-Pitch Linkage* |
| *200° Pitch* | *238° Pitch* | *280° Pitch* | *360+° Pitch is possible* |
| *360° Tool Roll* | *360° Wrist* | *360° Wrist Roll* | *360° Wrist Roll* |
| *360° Wrist Roll* | *360° Tool Roll* | *360° Tool Roll* | *360° Tool Roll* |
| *Standard design.* | *Same as Clevis, but Pitch is at an angle.* | *Same as Multi-Pitch, but sealed housing.* | *Requires boot for sealing.* |

Table 3.2  Dexterity Ratings for Pitch-Yaw-Roll Wrists

| | | | | |
|---|---|---|---|---|
| Distributed Single Axis Structure | Single Center Cardan | Double Cardan Joint Structure | Compacted Double Cardan Joint Structure | Serpentine Multi-Center Cardan Joint Structure |
| 180° Pitch 180° Yaw 360° Roll Less effective Pitch or Yaw range due to separate radii of rotation. | 180° Pitch 90° Yaw 360° Roll Difficult to integrate Roll Axis. Limited Yaw range same as ball-and-socket. | 180° Pitch 180° Yaw Roll provided by adding motor or rotating joint. High stiffness through massive structure. | 180° Pitch 180° Yaw 360° Roll Same as Double Cardan, but more compact. Features integral Roll. | Cantilever 360° Pitch 360° Yaw is possible Greater pitch range, but poor precision due to complexity and length. |

**Conclusion**

This chapter began by examining the importance of wrist actuators to robotics. The structure and kinematics of the human wrist provide a model and point of reference for robot wrist design. This understanding is central to designing robotic counterparts. Robot wrist actuator terminology and kinematic characteristics of the two primary types and their applications were given. The negative effect of singularity on wrist performance by causing detours in the path of the wrist was detailed. The advantages of singularity-free designs include greater efficiency and enhanced safety.

The first robot wrist design was sophisticated by any standard, no doubt owing to its base on the human wrist. Two established and proven designs were examined as reference points of design complexity. Recent NASA designs for use on multimillion dollar space robots continue to display a misunderstanding of dexterity. The evolution of the Omni-Wrist, the first high precision singularity-free design, was presented to give the reader a behind the scenes look at the creation of a wrist design.

As stated at the onset of this chapter, the primary function of the wrist is to orient the endeffector into a position or through a path, whether it be a spotwelder or spray painter. Wrist dexterity is ultimately determined by the endeffector's motion requirements. Of all endeffectors it is hands that we find the most fascinating and challenging, and it is the hand to which we now turn.

# 4

# *Hands*

*I will praise thee for I am fearfully and wonderfully made.*
Psalm 139:14

The hand can be described in mechanical terms as a differentiated endeffector. Grasping, manipulating, and pushing are just some of its capabilities. Each finger is a complex design that may offer four degrees-of-freedom, the ultimate in flexibility. Because of its complexity, its location on the arm, and its applications, the hand is also extremely dependent on the human body's ability to lubricate and repair itself. With each finger as complex as many robot arms, the hand multiplies control issues. Even more than the wrist, the hand's capabilities set a highly visible benchmark for robotic sophistication. It is indeed a great challenge for designers who have to work with inorganic materials to achieve the hand's functionality. The hand's joints and tendons present challenges in wear resistance and flexibility that often cause the subtle kinematics to be overlooked.

The popularity of designing and building robot hands is evidenced by the large number of universities and research organizations that have hands named after them. Almost every university had developed a hand by the mid-1980s, even though the subtle kinematics and drive sources were for the most part unresolved. A selection of the leading designs is presented with this chapter, ending with my Omni-Hand, which is in part a synthesis of many concepts but offers significant advances in kinematics and drive sources.

## The Human Hand

Composed of bones, muscle, cartilage and tendons, the human hand has fascinated both artists and scientists for centuries (Figure 4.1). Connected to the wrist by the palm, the human hand has a total of twenty degrees-of-freedom (four per digit) and is driven by approximately 40 muscles. Some muscles are located in the hand, but the majority of heavy lifting muscles are in the forearm and connect to the joints in the hand through tendons. Muscles in the hand provide the finely coordinated motion needed for grasping or manipulating. The hand's skeletal structure consists of the carpal and metacarpal bones that create the structure of the palm and fingers. Ball-and-socket type head knuckles permit adduction and abduction of the fingers (yaw motion). The muscles for driving the fingers' yaw motion are in the palm while muscles located in the forearm merge into tendons that drive the fingers' pitch motion. The palm is flexible and is composed of the four metacarpal bones held together by tendons and muscles which also allow flexure. The thumb is nearly identical to the fingers, but is mounted lower on the palm, and thereby facilitates grasping, adduction, and abduction. The head knuckle of the thumb, like those of the fingers, is also capable of circumduction, a circular motion created by pitch and yaw motion of the knuckle. Opposition, the ability of the thumb to touch the other fingers, is one of the key features

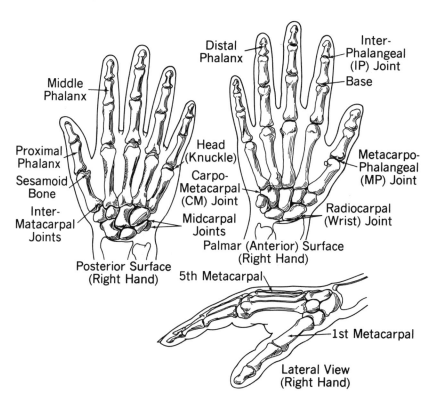

*Figure 4.1  Bone and Muscle Structure of the Human Hand*

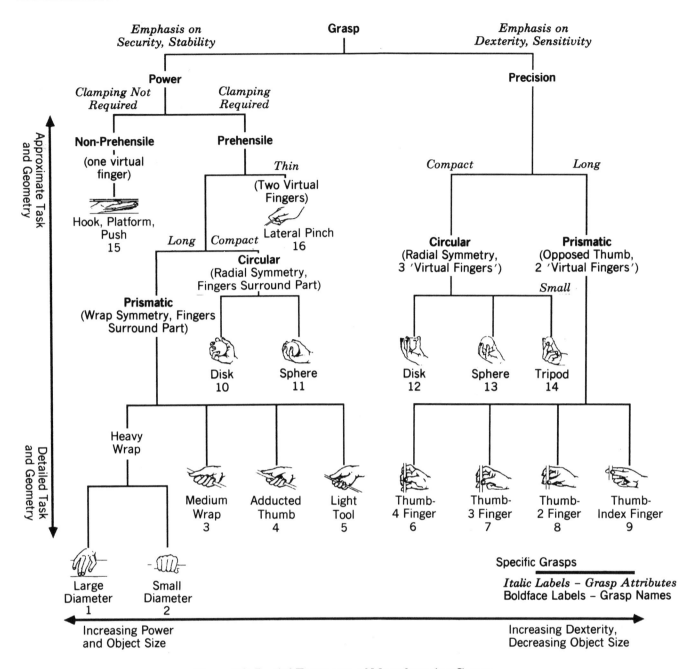

*Figure 4.2 Partial Taxonomy of Manufacturing Grasps*

of Homo sapiens that separates us from most other mammals and is essential for tool using. Figure 4.2 shows a taxonomy of how the human hand grasps and illustrates the human hand's great versatility. The kinematics of the hand are shown in Figure 4.3. and its motions in Figure 4.4. Disregarding the wrist, which was described in the previous chapter, the hand consists of a flexible palm which may be cupped to aid in opposition of the fingers. This cupping action becomes essential when grasping. The head knuckles are pitch-yaw ball-and-socket joints with limited passive roll capability. The upper two knuckles have one degree-of-freedom each. Capable of wonderfully quick motions, the human fingers are capable of tapping at rates approaching 10 Hz.

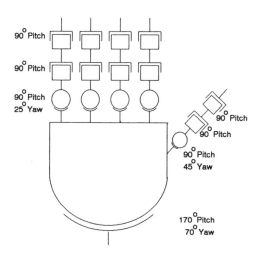

*Figure 4.3  Kinematics of the Human Hand*

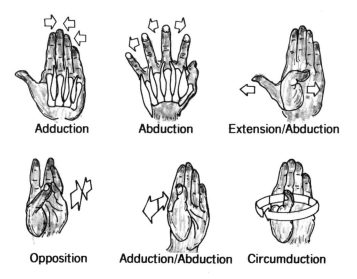

*Figure 4.4  Human Hand Kinesiology*

## The Musician's Hand

Before beginning on robot hands, a word should be said about the 18th century Jacquet-Droz creation known as the Musician. She is able to play a variety of songs that are still enjoyed to this day (Figure 4.5).

The Musician possesses two articulate five digit hands and plays an organ. The hands are powered and controlled by pinned cylinders in the Musician's body (Figure 4.6). Linkages pass through the hollow upper arm to the elbow where additional linkages permit the forearm to pivot. The fingers are bent to project the illusion of suppleness.

*Figure 4.5 The Musician*

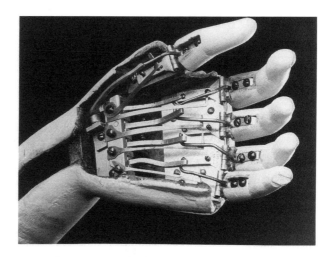

*Figure 4.6  The Musician's Hand*

A brass support screwed to the upper part of the hand receives the hinge on which the cranks pivot [Figure 4.7].  The crank is connected to the rod in the forearm by a stud, while the second part is formed into a brass shank, which just presses on a steel wire set in motion, the crank pivots and presses on the steel wire of the finger, while a return spring makes it take up its initial position when the rod is no longer moving (Scriptar-F.M. Ricci, 1974).

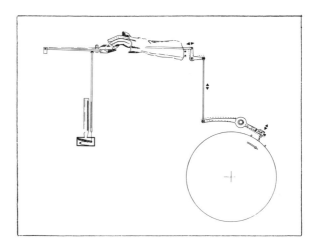

*Figure 4.7  Detail of the Musician's Finger*

Here we have an example of a programmable (albeit mechanically) pair of articulate fingers driven remotely and connected to pivotable arms. It is humbling to see this execution and practical application so far ahead of its time.

## Robot Hands

Robotic hands began with the Argonne teleoperator parallel gripper designs. Many design morphologies have been sought and typically reflect the limits of materials or the designer. Although they may be better equipped than the human hand to perform certain dedicated tasks, they fall short in the general purpose applications. This is a common problem with robot designers who become enamored with their own design, justifying it by one feature or application alone. Since the beginning of modern robotics, the Holy Grail of a truly anthropomorphic hand has been sought but not found.

As industrial grippers, the hands of the robot have played an important part in the advancement of industrial robots. Two basic designs have come to dominate the industrial market. Pivoting finger grippers (Figure 4.8) are usually actuated by an air cylinder in the base of the gripper. The fingers pivot outward and inward in an arc. Parallel jaw or finger grippers are also named from the action of the fingers (Stauffer 1987a). The fingers are usually actuated by an air cylinder in the base of the gripper but move in a common plane and remain parallel to each other. (See Pham and Heginbotham 1986 and Kato 1982 for exhaustive studies of these two types.) Two approaches are used to cope with the low dexterity of these two types of grippers (Figure 4.9). The turret gripper uses multiple customized grippers mounted on a rotating turret that indexes the appropri-

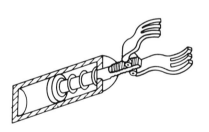
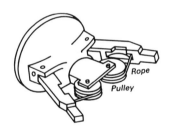
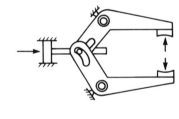
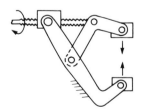
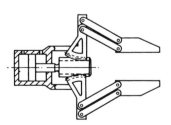
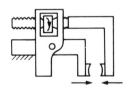

*Figure 4.8 Various Forms of Pivoting Jaw (top row)*
*and Parallel Jaw Gripper (bottom row)*

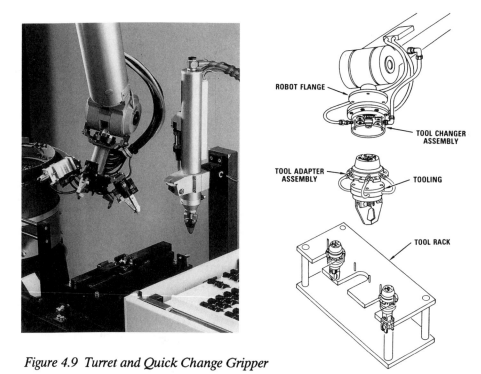

ROBOT FLANGE

TOOL CHANGER ASSEMBLY

TOOL ADAPTER ASSEMBLY

TOOLING

TOOL RACK

*Figure 4.9  Turret and Quick Change Gripper*

ate gripper into place. The advantage of this design is that high speed and simple gripper changeover is achieved by simply rotating the turret. The quick change gripper, like the turret gripper, uses multiple customized grippers stored in a rack or other dispenser from which the robot selects the appropriate gripper as job requirements change. Often employed with larger, more complicated grippers, it allows for a smaller endeffector than the turret gripper. Internal porting of air and electrical connectors is used to transfer control and provide communication with its sensors. Compliant interfaces are often added to cope with the low dexterity of the fingers and wrist (Figure 4.10).

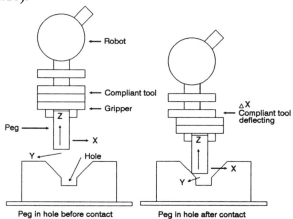

*Figure 4.10  A Compliant Tool Interface*

## General Electric Handyman Hand

Articulated multijointed telerobot hands can be traced back to "Handyman," a two armed electrohydraulic teleoperator developed by Ralph Mosher at the General Electric Corporation in 1960 (Figure 4.11). Two fingers connect to the wrist pitch axis and the forearm provides the wrist roll. The fingers have five pivoting segments that provide three degrees-of-freedom. Joints 4 and 5 work together to produce a compound articulation. Each finger effectively has three degrees of freedom. A two degrees-of-freedom wrist is controlled by the operator, who was "wearing" the slave apparatus. A single command causes the fingers to grasp an object in a prehensile manner (Vertut and Coiffet 1986).

Since 1969 work has progressed primarily in the United States and Japan. Most robot hands have two fingers and a thumb. So called anthropomorphic robot hand designs, typically, consist of a number of pivoting, hinge-like joints making up the finger joints. The head knuckle consists of two separate rotary joints, thereby complicating the design and computer control. Because the axes of the two rotary joint knuckles are distended they cannot be coupled, and this makes computer control computations more time consuming.

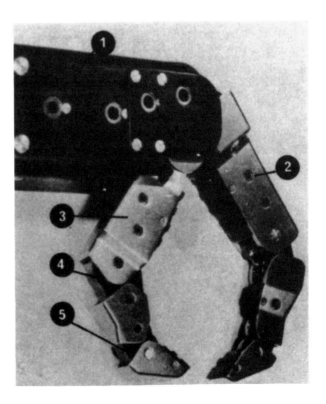

*Figure 4.11 Handyman by General Electric*

Despite Handman's success, engineers have never overcome the limitations of tendon based systems. Further tendon research is needed to make tendon powered robotic hands practical, although they may never compete with the reliability of direct drive hands. Since their inception, steel tendons and cables have been used to drive teleoperators in nuclear handling applications. Unfortunately, steel tendons have a tendency to break unpredictably and are difficult to replace remotely. The interface connecting the tendons to the drivers and linkages is a chronic failure point because of the abrupt transition from soft to hard materials. Progress on improved tendon material has been made at the University of Utah. Dacron-Kevlar blends woven together have been used successfully in anthropomorphic robot hands (Jacobsen 1986). The current trend is to drive the joints directly, as in the Odetics Hand and Omni-Hand discussed below. Development of these designs poses special challenges in kinematics, structure and mechanical connections.

## Salisbury Hand

Originally called the Stanford/JPL (Jet Propulsion Laboratory) Hand, the Salisbury Hand was designed by Kenneth J. Salisbury as part of his doctoral research. The hand weighs 2.4 Lb (1.1 Kg) and the drive assembly weighs 12.1 Lb (5.5 Kg). Output force is 9.9 Lb (4.5 Kg) for up to two minutes. The unit has been sold primarily to university and corporate research departments for laboratory demonstrations (Figure 4.12). In an ironic conflict of complexity versus simplicity, each three degrees-of-freedom finger has no less than four Teflon coated control cables that slide in Teflon lined conduits, yet the fingers are modular to reduce the number of parts. Finger position information is produced by strain gauge sensors and motor position sensors located behind each proximal joint. The tension signal is translated into a joint torque signal used to close the servo loop. Fingertips feature a highly compliant rubber-like material that provides friction and "give" for a secure grip (Mason & Salisbury 1985).

An actuator pack of 12 samarium-cobalt LO-COG DC servo motors with 25:1 speed reducers power cables that drive the fingers (Figure 4.13). Each of the three fingers is composed of a double jointed head knuckle that provides the joint with plus or minus 90 degrees of pitch and yaw motion. An additional knuckle above the head knuckle has a range of plus or minus 135 degrees. This increased range compensates for the absence of the third knuckle found in the human finger, but unlike the human head knuckle, the pitch and yaw axes do not intersect in the head knuckle. Advantages of this include the simplicity of modular fingers and the lower cost of parts. Dexterity, however, is relatively poor because the head knuckles stack one axis above the other. The flexible cable drivetrain is less reliable than a direct drive system. In addition to being unreliable, push/pull cables have inherently limited power transmission capability and are difficult to route through the wrist.

Some of the control and utilization issues Salisbury grappled with are addressed through a modified PUMA controller for the electronic interface. Customized software for driving fingers and interacting with sensors provides data fusion. Salisbury predicts the next generation of motor control will be DSP based (digital signal processor), with specialized processors for number crunching, with some form of DSP adopted for motor control. The majority of Salisbury control work is in the area of "fingertip prehension." The object is already grasped in the fingertips with the fingers imparting motion to it. New control areas being explored include: "controlled slip manipulation" which allows momentary slippage or loss of contact between the fingers and the object. Broader research efforts also include coordinating the fingers to achieve secure grasps and simple assembly/disassembly operations as well as tool handling (Pellerin 1992).

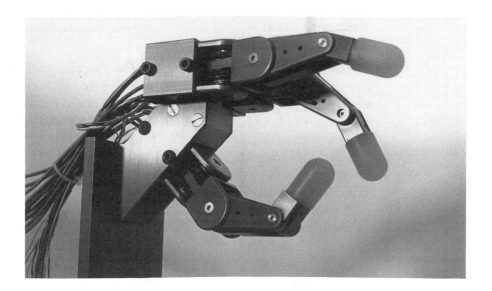

Figure 4.12  Stanford/JPL Hand (Salisbury Hand)

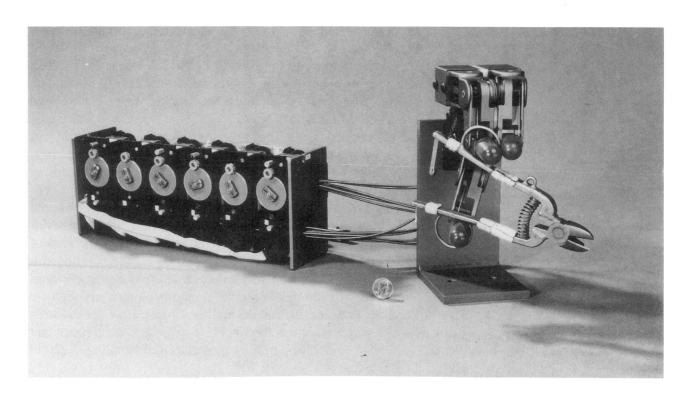

Figure 4.13  Stanford/JPL Hand (Salisbury Hand) Drivetrain

### Victory Enterprises Hand

Three fingers and an opposable thumb provide this hand (Figure 4.14) with an anthropomorphic configuration (Ladendorf 1984). It is connected to a four degrees-of-freedom arm. Although this hand uses conventional tendon technology the drivetrain is of interest because of the unique technique of multiaxes motion (Petersen 1984).

Two high helix leadscrews continuously rotating in opposite directions provide a bidirectional "power bus" which extends the length of the system (Figure 4.15). "Actuator modules" contain a pair of high helix leadscrew nuts coupled to electromagnetic brakes. When the brakes are energized the freewheeling nut becomes engaged, thus moving the actuator module along the lead screw. Force may be varied by controlling the current to the brake, but speed control seems difficult with this method.

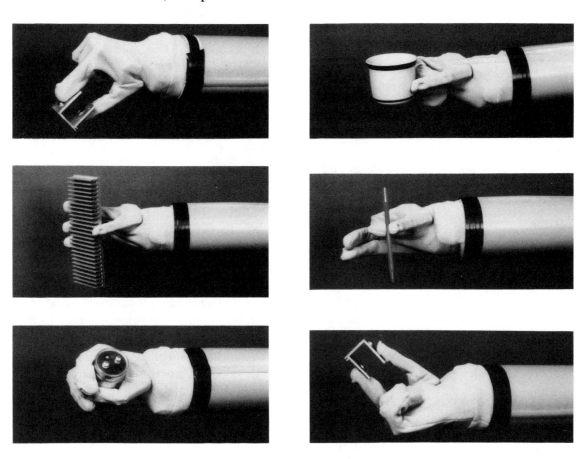

*Figure 4.14 Victory Enterprises Hand*

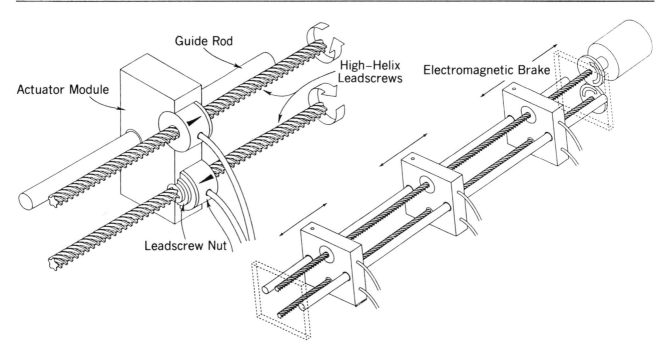

*Figure 4.15 Drivetrain of Victory Hand*

## Crew and Thermal Systems Division (CTSD) I Hand

NASA has a long tradition of funding hand designs. Some early designs included the Multiple Prehension Manipulator system for Skylab (Skinner 1975a, 1975b) and End-effector Device (Clark & Johnson). Developed by Johnson Space Center's Crew and Thermal Systems Division (CTSD) for deployment on Extravehicular Autonomous Robot (EVAR), discussed in Chapter 6, this hand has a single degree-of-freedom (**Figure 4.16**). Each finger contains three sections connected by joints that are coupled together by direct linkages. The push/pull action created by the rod inside the proximal finger section causes the other sections to move in unison. Three fingers 120 degrees apart allow closure for picking up objects. An automatic fingertip trap mechanism ensures secure gripping. As the finger begins to close, the distal finger section bends around the object being grasped. A single motor drives a takeup/payout drum via a small plastic belt. Each finger has the cable running though a small pulley so that as the cable is wound up, each finger is drawn in. If one finger stops, the others continue to close in a simple, differential manner. Structurally simple, the main feature of this design is its ability to lend itself to straightforward, dedicated applications. A relatively small number of parts provide multiple functions; for example, the drum not only promotes simplicity but also enhances the reliability. However, functionality is limited to simply grappling objects, not manipulating them.

## CTSD II Robotic Hand and Wrist System

An advanced version of the CTSD I, this hand is equipped with a simple wrist (Figure 4.17). Each finger is independently controlled by its own DC motor for improved dexterity that represents an early attempt at modularity. To visualize the performance improvement this produced, imagine what your hands would be like if all of your fingers were forced to move together in unison. The CTSD II's fingers are arranged in a two opposing one configuration that provides parallel grasping surfaces (Hartsfield 1988). Lifting circuitry in the central electronics limits the maximum amount of force each finger can exert. A linkage in each finger transfers the base knuckle motion to the one above it. The wrist is of the pitch-roll type, with motors driving bevel gear pairs in a classic differential arrangement (Devol 1967). Leading the way in robot hand sensor technology, the hand features grip force measurement devices. Piezoresistive film sensors are applied to the finger surfaces to provide tactile feedback that gives the hand a sense of touch. Silicon pads cover the tactile sensors for protection and provide a compliant surface for secure grasping. Infrared sensors have also been added for object detection within the fingers, while strain gauges applied to structural members provide a redundancy to this object detection. The same sensor technology has been applied to Johnson Space Center's MIT/Utah hand in a demonstration in which the hand catches a ball by closing its fingers after sensing the ball is within its grasp (Hess 1990). A conventional two degrees-of-freedom wrist manipulates the hand. Motors located in the forearm drive bevel gears for a differential pitch/yaw action.

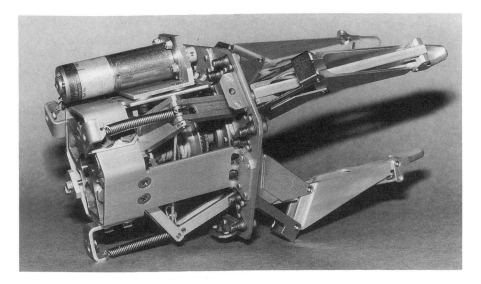

*Figure 4.16  CTSD I Hand*

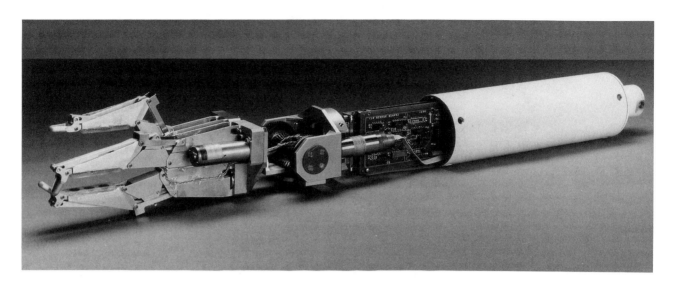

*Figure 4.17  CTSD II Hand*

## Hitachi Hand

The Hitachi Ltd. hand (Figure 4.18), with its radical shape memory alloy (SMA) metal actuation technology, caused a mild sensation in the robotics world when it premiered in 1984 (Hartley 1984a). Its overall length is 27.5" (70 cm), including the forearm, and it weighs 9.9 Lbs (4.5 kg) while having a 4.4 Lbs (2 kg) load capacity. Claiming a 10:1 reduction in weight over previous designs, its power-to-weight ratio and compactness are its chief attributes. It has three four-jointed fingers and a thumb with twelve actuators per finger. A pitch-yaw wrist is also included. The Hitachi Hand uses a large number of very thin SMA wires that can be heated and cooled rapidly. Each finger has a number of SMA wires .02 mm in diameter that are set around the tube housings of the spring actuators. For the wrist, each axis has four SMA wires .035 mm diameter set around pulleys. A fast 90 degrees per second of joint travel has been achieved. The SMA wire, when heated by passing electric current through it, reacts by contracting against the force of the spring (Figure 4.19). The joint angle is varied by modulating the current passing through the SMA wire. The current is proportional to the error between the microcomputer controlled reference angle and the angle of the joint detected by a potentiometer on each actuator (Yoshiuki 1984).

Despite its high speed and load capacity, the longevity and ruggedness of the SMA actuators are questionable. SMA, like any metal that is repeatedly flexed, eventually suffers fatigue and fails. However, it is interesting from the standpoint of being a radical approach to the classic robot hand power problem-getting power to the fingers while maintaining a compact and lightweight wrist design. Applications predicted for this hand are for complicated maintenance work in hostile environments. Beyond this, medical micromanipulators, semiconductor handling robots, and undersea operations were foreseen.

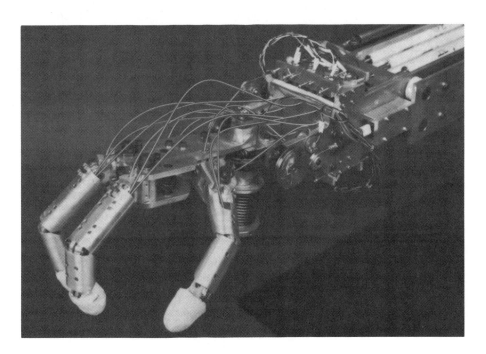

*Figure 4.18  Hitachi Hand*

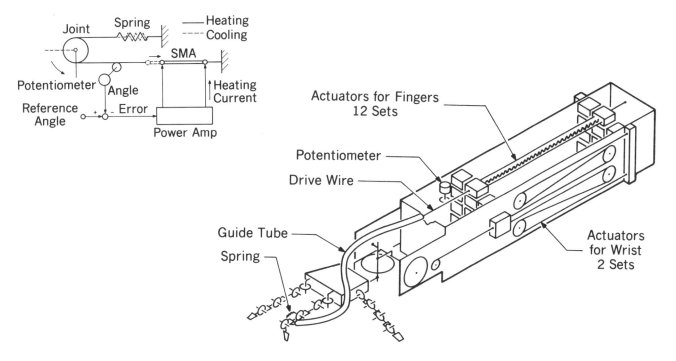

*Figure 4.19  Hitachi Hand Showing Actuators and Guide Tube*

## Utah/MIT Hand

One of the pioneer designs from the mid-1980s is the Utah/MIT Dexterous Robotic Hand (Figure 4.20). The Utah/MIT hand was developed to perform laboratory research on grasping and finger manipulation, and design creativity shown in the feedback sensors and controls, but less in the physical design. A previous tendon driven design (Mullen 1972) had three fingers and a thumb that provided a total of sixteen degrees-of-freedom (four per digit).

The Utah/MIT Hand closely copies the outward appearance of the human hand and is modular in design; each finger is identical. However, cabling is very complex and requires a separate, articulated arm-like frame because each joint is controlled by a pair of antagonistic tendons. The head knuckles are not like the human ball-and-socket design that has two intersecting perpendicular axes. Instead, a non-anthropomorphic roll/pitch is used which lacks the circumduction capability of human head knuckles (Figure 4.21). Tip force, the force exerted at the finger tip, is 7 Lbs (3.18 Kg) and frequency components exceed 20 Hertz. This high dynamic performance is characteristic of many cable operated systems because tendons make low mass designs possible. Although tendon life cycle continues to be improved, the reliability of cables is poor in comparison with the reliability of solid steel shafts or push/pull rods. Unpredictable failures occur at the transition point where, unlike human tendons, an abrupt transition from flexible to inflexible material is made.

Powered by 32 50-to-100 psi pneumatic double acting glass cylinders (Figure 4.22), inside each knuckle is a linear Hall effect sensor that provides joint angle information. Hall effect sensors also monitor tendon tension in the wrist and are used in the finger knuckle joints for position sensing. Design allowances were made for tactile sensor wiring. The wrist is a conventional unpowered U-joint design (Figure 4.23). Range of motion is less than that of the human wrist with pitch 90 degrees, yaw 30 degrees, roll 270 degrees.

The tendon drive system was an obvious short term design convenience that cascaded negative effects into the design. Control is difficult because the tendons are compliant, while wrist range and finger kinematics are compromised. tendons made the design nightmarishly complicated; friction requires an elaborate system of 288 pulleys (Jacobsen 1984a, 1984b, 1988).

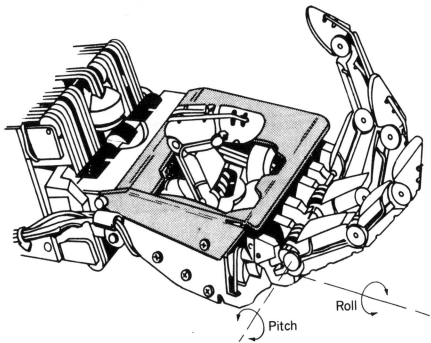

*Figure 4.20 Utah/MIT Hand*

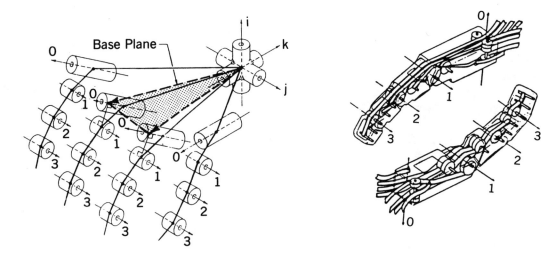

*Figure 4.21  Kinematics and Tendons of Utah/MIT Hand*

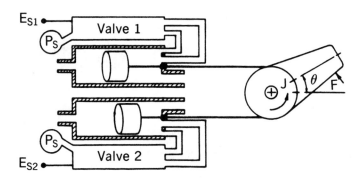

*Figure 4.22  Utah/MIT Hand Double Acting Air Cylinders*

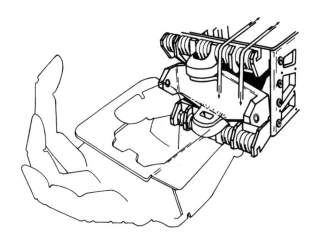

*Figure 4.23  Wrist from Utah/MIT Hand*

## Jameson Hand

The Jameson Hand, a robotic version of the direct link prehensor (Hartsfield 1988, Jameson 1987) and Utah/MIT Hand, features a similar complicated array of tendons. However, the tendons are routed through three conduits that protect them from entanglement and damage. The conduits make for a more rugged and practical design (Figure 4.24), but they restrict the wrist's range of motion. Finger and wrist motors are located in the forearm. A standard bevel gear type wrist is used which was developed for the nuclear industry. The sensing system borrows from the Utah/MIT Hand's adaptive grasping development.

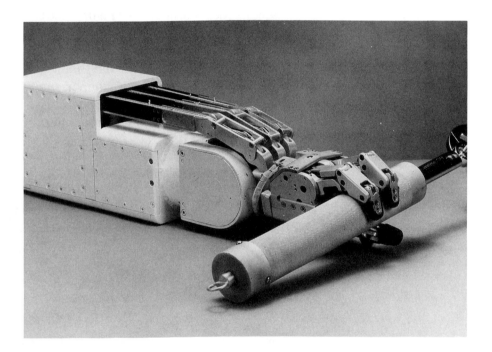

*Figure 4.24 Jameson Hand Design*

## Belgrade/USC Hand

The Belgrade/USC Hand (Figure 4.25) was developed jointly by the Electrical Engineering and Computer Science Departments at the University of Southern California and the Electrical Engineering Department at the University of Belgrade, Belgrade, Yugoslavia (Venkataraman and Iberall 1990). A general anthropomorphic design, it has four fingers and a thumb and is capable of holding 5 Lb (2.2 Kg). Grasping is accomplished by "synergistic flexion" of all joints in unison. Four motors drive the hand; two for the fingers and two for the thumb. Closing the hand from full extension takes two seconds, while thumb closure is twice as fast as the fingers. Each finger motor drives two adjacent fingers for pitch through a screw/nut drive. The fingers are compliant in the yaw axis by virtue of the unique rocker arm design (Figure 4.26) that automatically increases finger travel if no object is encountered. The opposable thumb rotates in yaw 120 degrees from full extension to the center of the palm. One thumb motor provides swivel for the thumb's opposition motion and the other provides prehension in a manner similar to the fingers. Prehension is achieved by the upper knuckles of the fingers and thumb, which are powered by a system of linkages that transmit the lower joint's motion to the one above it.

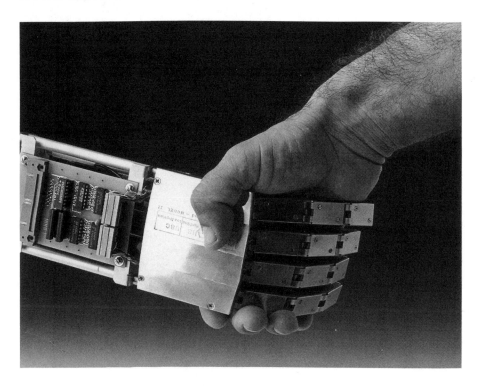

*Figure 4.25  Belgrade/USC Hand*

Absolute position linear potentiometers are used as position transducers and allow the user to monitor a constantly available DC voltage representative of the "Finger-Thumb" flexion-extension position. Twelve force sensors, two on each digit and two located on the palm, provide a logarithmic indicator of the forces exerted on the hand. Each sensor's output voltage is monitored by the controlling computer, which allows for customized programming of functions. A rubber glove is also available to protect the hand from particulate matter. Future improvements include incorporation of slippage feedback into the control system that would be able to detect "unstable grasps." Feedback control automatically adjusts the hand's shape to stop slippage. A broader area of research is increasing intelligence in hand shape control software. Neural networks that "learn" grasp posture relative to object geometry already have been achieved in some simple applications.

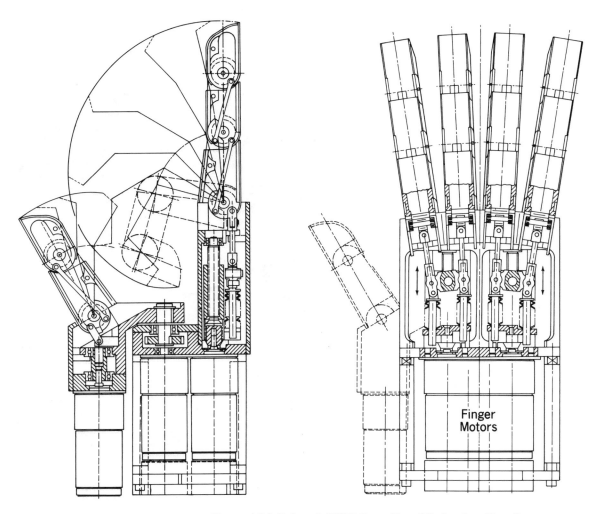

*Figure 4.26 Belgrade/USC Compliant Mechanism Detail*

## Odetics Hand

The Odetics Hand is a powerful human sized hand, self-contained in a compact package. An early prototype is shown in Figures 4.27 and 4.28. A modified unit was delivered to NASA Johnson Space center in 1991, but dexterity was less than the earlier designs. Other than publicity photos, little material has been released on this design. A two speed transmission and clutch mechanism coupled with force sensing provide the fingers with a dual speed mode: fast as the finger closes in on its object and slow after it achieves contact. Each finger has a drive motor and solenoid. When the solenoid is energized the finger base is rotated to create different grasping positions, when locked the upper finger knuckles are actuated. Described by its inventor Steve Bartholet as both hand-like and tool-like, it represents the opposite extreme from the MIT/Utah design, a machine or claw-like structure in contrast to an anthropomorphic design (Stauffer 1987b). Dexterity is quite low because of the simple kinematics. This is a good example of the difference between simple and complex joint kinematics. Simple kinematics may be packaged by the structure itself, whereas universal type motion requires a flexible covering material rather than a rigid enclosure.

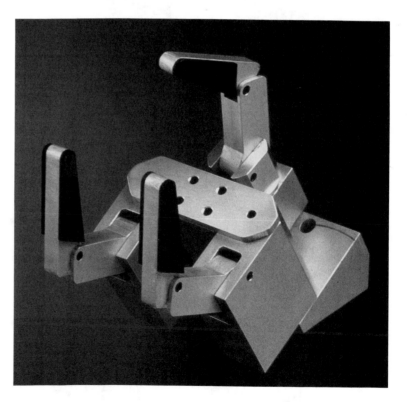

*Figure 4.27  Odetics Hand*

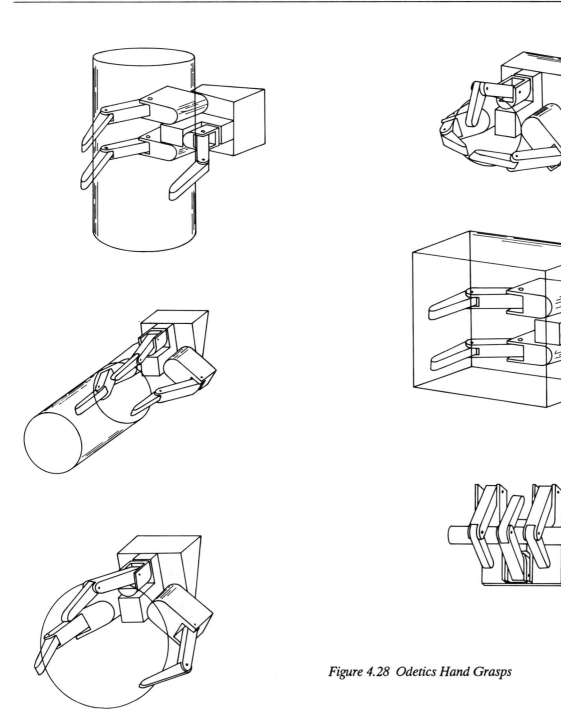

*Figure 4.28 Odetics Hand Grasps*

## Sarcos Hand

An outgrowth of the MIT/Utah hand project designs, the Sarcos Hand (Figure 4.29) is a three degrees-of-freedom hand. Faced with the maddening complexity of closely copying the human hand, this design compromises on degrees-of-freedom and dexterity in return for a simplified direct drive (Korane 1991, Jacobsen 1991). The elimination of tendons also made possible a higher angulation single center pitch-yaw-roll wrist. Wrist pitch is 90 degrees, yaw 90 and roll 180. Note the significant distance between the center of the wrist and the center of the hand. This is detrimental to dexterity. Powered by 3000 psi hydraulic pressure, this hydraulic hand has a 50 Lbs (22.7 Kg) load capacity. A miniature double acting hydraulic cylinder and hydraulic rotary vane actuators power the thumb's head knuckle to 90 degrees of pitch and yaw. The outside finger is capable of 45 degrees of rotation while the middle finger is fixed. Hook-like in appearance, it recalls prosthetic hands that use hooks to pick up objects. The middle finger provides the absolute reference point required by the Hall effect sensor used to measure the position of the moveable fingers. Like the Odetics Hand, a seemingly impressive array of grasps may be achieved (Figure 4.30). The reality is that the human hand's kinematics and degrees-of-freedom are still unrivaled, as evidenced by Figure 4.2.

*Figure 4.29  Sarcos Hydraulic Hand*

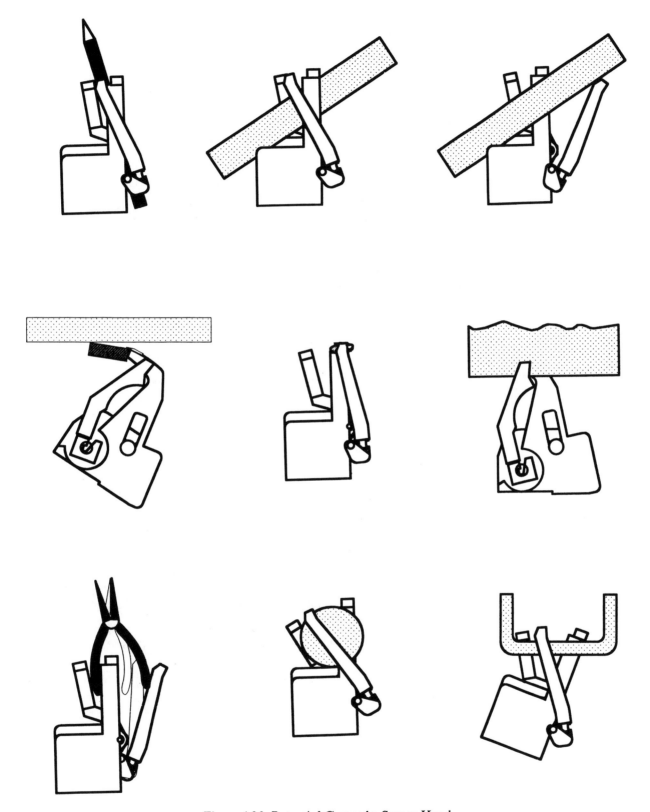

*Figure 4.30  Potential Grasps by Sarcos Hand*

### Omni-Hand

Ross-Hime Designs' anthropomorphic robot hand, known commercially as the Omni-Hand (Figure 4.31), improves on the previous designs by reproducing the motions of the human fingers and wrist in a more life-like and dextrous manner. Funded under NASA contracts NAS8-37638 and NAS8-38417, it received the 1993 NASA Technology Utilization Award.

Basically a hybrid, it has the ruggedness of the Odetics Hand while exceeding the dexterity of the MIT/Utah hand. The fingers are modular like the Salisbury Hand. As in the human hand, the adduction and abduction (yaw) of the finger is direct drive powered by electric linear actuators in the palm. Additional characteristics include high power, rugged design, anthropomorphic kinematics, and nonbackdriveable fail-safe joints, all of which are important for telerobot control.

Used as both a wrist and knuckle, the ball-and-socket head knuckle features circumduction identical to the human hand, a design feature vital for creating an opposable thumb (Rosheim 1989a, 1989d). This pitch/yaw motion produces dexterity and simplified computer control not found in other robotic hands. The head knuckle socket is biased to allow a greater range of motion in one direction, as in the fingers of the human hand. The joints also have built-in stops to prevent backward hyperextension. Modular in design, all the fingers operate in the same way and are interchangeable, bolt-on modules. Actuation of the head knuckle in pitch/yaw motion is comparable in range to the human finger's head knuckle.

As in the Belgrade/USC Hand, linear actuators were seen as the most practical method of providing high power in a compact package. An important advancement was the concept of modular or building block linear actuators that could be used in multiple locations. Development of the linear electric linear actuators liberated the design from dependence on tendons and freed the design from their limitations. The upper finger linkage is directly driven by a Minnac (the name being derived from miniature actuator) that causes the spur gearing in the uppermost joint to reproduce the motion of the lower joint. This action is similar to the prehensile effect in the human finger, where the uppermost joint mimics the lower joint's angle.

In head knuckle pitch motion, the Minnacs power a sector spur gear, which in turn drives a ring sector connected to the cup. This sector is held in the ball groove by two pivotal pins, the sector sliding across the ball. The head knuckle consists of two half-balls mated together; contained in this ball is a second ball with a pin press fitted into it. The purpose of this ball is to retain the finger socket on the head knuckle ball without restricting its two axes motion. Yaw motion of the head knuckle is produced simply by an arm protruding out of the finger socket connected to a modified

*Figure 4.31  Omni-Hand - Front View*

U-joint. The U-joint is necessary to decouple the two axes motion of the knuckle so that it may be driven by a linear drive.

The wrist (Figure 4.32) is simply an enlarged version of the head knuckle with its pitch and yaw actuators mounted on the forearm through pivotal bearings. Wrist pitch is 110 degrees with wrist yaw at 70 degrees. Wrist roll may be added by simply interposing a motor between the elbow and the forearm section.

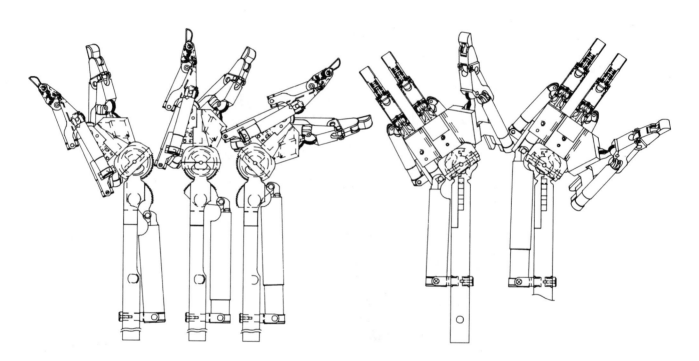

*Figure 4.32 Wrist from Omni-Hand*

A high level of mechanical and electrical modularity of the hand was accomplished. Each finger is an interchangeable bolt-on unit for ease of maintenance and minimal cost, like the Salisbury Hand. Electrical quick disconnects for motor power, position sensing, and tactile sensing were achieved. This greatly enhances maintenance and simplifies logistics by minimizing parts inventory. The first Omni-Hand was built as a test bed with the second unit incorporating the first's desirable features.

Dry lubrication techniques, such as Dicronite and ion implantation, to coat the Omni-Hand's spherical bearing surfaces, gear teeth, and screws were researched and successfully applied to the design. The complexity required a low maintenance lubricant to coat the wrist ball-and-socket, knuckle ball-and-socket, Minnac threads, sector gears and sector gear

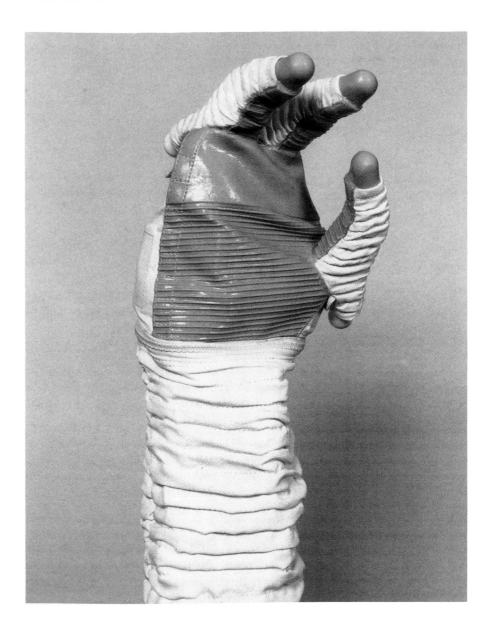

*Figure 4.33 Space Glove for Omni-Hand*

pivots. Self-lubricating plastics such as Delrin AF were used in the Minnac screws to reduce friction.

Overlooked by the majority of hand designers, the complex, almost organic nature of the Omni-Hand mechanical and electrical systems required protection against debris, collision, or other environmental hazards. Like the hand's human skin, a glove is necessary to protect the intricate parts of the Omni-Hand.

Many of the challenges such as flexibility and abrasion resistance posed in the glove design had already been addressed by the makers of the NASA astronauts' space suits, so they were subcontracted to construct two prototype gloves. Requirements were minimal; the gloves were to be high flexible and provide protection from the environment, but a close working relationship ensured that the design and work was on target.

The glove's exterior is made of a Teflon coated Kevlar/Nomex blend (Figure 4.33) known as "Ortho Fabric," which is the external fabric of space suits. Ortho Fabric panels are sewn to an inner layer of elastic spandex which creates convolutions to allow flexing while still maintaining a "clean" and compact exterior. Flaps were located to allow access to the hand's critical internal components. Flexible Teflon guards were sewn into strategic areas in both hands to prevent abrasion of the fabric by the hand mechanism. A zipper at the midpoint of the wrist facilitates installation or removal of both gloves. Textured elastomer surfaces and silicon rubber finger tips were added to the palm and inner finger surfaces to increase tactility and gripping. In the second glove, sealing was improved by designing spandex panels over the finger motor areas.

Development of linear electric linear actuators for optimizing effective force within the small envelope of knuckle and wrist operation was critical to design success. In the Phase I design this required using miniature linear actuators. Dissatisfaction with conventional large and bulky linear actuator technology led to the new patented concept known as the Minnac (Trechsel 1992). Figures 4.34 and 4.35 show the Minnac that was developed for the fingers. A .63" (16 mm) diameter custom designed, 24V dc motor with integral encoder and gear head was also chosen for cost effectiveness and reliability. The output shaft connects to a split screw which is preloaded to eliminate backlash. Double bearings support the motor shaft on which the drive screws are mounted. The double bearings make the Minnac very rugged and insulate it from shock loads. Torque is increased through the 76:1 gear box and speed is increased through multistart threads on the screw. This simple concept of a slow high-ratio gear box driving a fast screw give the Minnac its high efficiency. Over 13 Lb (6 Kg) of vertical lift for each finger, and over 30 Lb (13.5 Kg) of vertical lift was achieved for the wrist. Approximate speed is .5" per second.

Improvements and optimization of the palm for finger placement, including the thumb, was a major design consideration. Palm developments also included the investigation of interchangeable flexible pad inserts on the inside of the palm to aid in gripping the workpiece. Two palm concepts were developed. The first produced a claw-like finger configuration with each finger connected to a detachable module. Experimentation proved that this design would be less useful for general grasping. The second, a more anthropomorphic design that drops the thumb down and makes opposition possible, was eventually adopted.

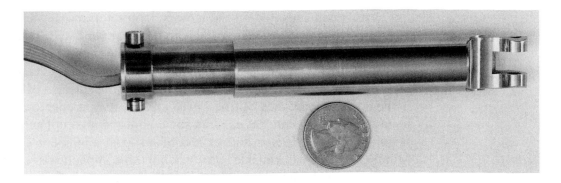

*Figure 4.34 Minnac Actuator*

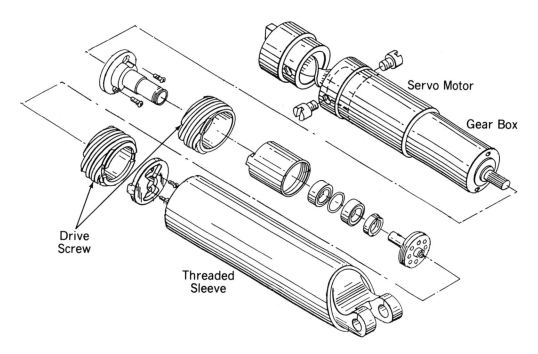

*Figure 4.35 Exploded View of Minnac*

Testing concentrated on the motors, motor servo loops and assembled linear actuators. Mechanically both the finger and wrist linear actuators performed spectacularly and load goals were exceeded. The assembled Omni-Hands were used as test fixtures for dynamic servo control and motion profile testing, and to check the stiffness of each joint in the pitch and yaw axes. Stiffness in all joints is comparable to the human hand's. The head knuckle exhibits some additional compliance when a finger is vertical. This is due to redundant axial alignment of the yaw Minnac and the head knuckle's yaw axis, but this compliance is comparable to that of the human head knuckle. Resolution, the repeatability and accuracy of the actuators, were tested under various loading arrangements. Velocity and acceleration testing with load indicated excellent performance for teleoperator or robotic control.

During development of the Omni-Hand the need was perceived for a highly flexible wire harness to provide electrical power and position information to the multiple electric linear actuators located on the wrist and hand. Secondly, a flexible wire harness was required for electronic tactile sensors located on the fingertips and palm. These two tasks were complicated by the anthropomorphic and therefore very flexible nature of the Omni-Hand. Multiple electric linear actuators above highly flexible joints required ingenious solutions to prevent binding, abrasion, or other undesirable damage to the harness. The glove and wiring concepts evolved together.

Routing the wires from the base of the forearm to the wrist was solved by an eyelet mounted on the side of the forearm. It allowed the bundle of wires to slide up and down in response to the wrist flexing. A high flexibility wire was selected for use. Wire harnesses for the fingers posed a greater challenge. The first wire harness approach for the linear actuators consisted of routing the wires over knuckles. Undesirable wire lengths resulted that would require elaborate takeup pulleys and springs, and would create a bulky and complex design. The second approach considered containing the wires in pockets and passageways inside the glove. Weaving the wires into the glove fabric also was considered, although this was dismissed as too complex and costly, not to mention making removal of the glove difficult, and potentially restricting maintenance of the hand's internal mechanisms. The third and final design, which was adopted, was to route the wires alongside the wrist to minimize bending (Figure 4.36). As the wires rise into the back of the palm, they spread out to the various lower linear actuators and attach to them through miniature connectors. The upper motor cables are guided by an eyelet integral to the palm, located between the two fingers (a similar arrangement could be added if more pairs of fingers are added). Small clips hold the wires to the sides of the finger knuckle. This prevents entanglement or undesirable abrasion, and minimizes stretching by holding the wires at the pivot points of the knuckles.

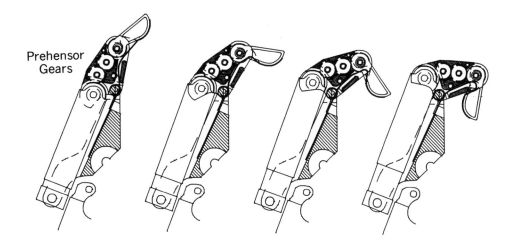

Prehensor
Gears

*Figure 4.36 Passage for Tactile Sensors*

The wire harness for the tactile sensors posed a greater challenge. A cable tied to the linear actuator wire bundle passes through the forearm eyelet and up the wrist. This cable connects to the circuit board mounted to the back of the palm. The circuit board has four connecters mounted to it, one for each of the three fingers and one for the palm sensor. The circuit board provides quick electrical disconnection of all sensors. It was foreseen that the tactile sensors would be subjected to high levels of wear and would require frequent replacement. Multiple tracing, ribbon-like flexible circuits terminating in sensor elements, connect to the circuit board connectors. From the circuit board they are routed to the head knuckle and are split apart to increase their flexibility at this pitch-yaw joint. A dogleg in the flex circuit allows it to pass under the linear actuator and over the head knuckle. The flex circuit then passes into a channel in the first finger segment between the head knuckle and the second knuckle, and continues through an internal passageway in the second finger segment (Figure 4.37). A small pocket at the base of the final segment accommodates any excess flex circuit caused by the pivoting of the joint. Because tendons were not used the passage was possible.

Research and development of the Omni-Hand was successful. Multilevel integration of kinematics, structure, wiring, and enclosure was accomplished through multilevel thinking and research into past designs. The first high dexterity, modular, electromechanical Omni-Hand was produced for NASA space application.

The Omni-Hand provides a synthesis of the best features from the preceding design approaches: an anthropomorphic configuration with direct drive. The creation of miniature muscles first and building a me-

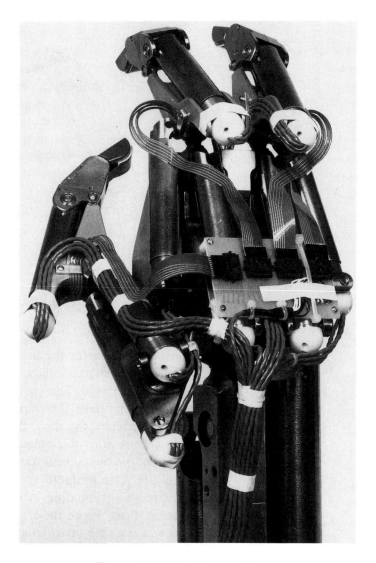

*Figure 4.37  Wire Harness Concept*

chanical design around these produced benefits in modularity and cable routing. Development of directly driven dextrous joints that closely imitate the human hand's knuckles was a major accomplishment. A follow on contract with NASA Johnson Space Center focuses on increasing the anthropomorphism; reducing weight; increasing speed; adding on anthropomorphic flexible palm to aid in gripping; and miniaturization to increase compactness and decrease size. Future enhancements will include additional fingers; development of reliable miniature brushless motors; and utilization and integration of all tactile sensors in full array mode.

**Conclusion**

We began by examining the human hand's kinematics and structure as a reference for designing robot hands. A brief, and somewhat humbling, glimpse at the 18th century Rococo era showed that our generation was not the first to work this problem.

The evolution of robot hands, in comparison to arms, has only recently begun. Simple open-close, quick change, and turret grippers are common in industry. Hostile environments and hazardous material handling continues to provide impetus for versatile anthropomorphic hand designs. The greatest mechanical design difficulty is posed by the hand's multiple dextrous joints in a compact package with a daunting drivetrain. Designs such as the MIT/Utah Hand dead-ended because they attempted to emulate nature too closely. Designers have focused on the drivetrain in an effort to minimize complexity. Early efforts to improve or eliminate drivetrains include the Hitachi Hand and the Belgrade/USC Hand. Hybrid designs like the Odetics and Sarcos Hands have a future because they use a minimum of components but achieve some of the functions of the human hand. The grand challenge to design a hand that is both rugged and anthropomorphic continues to attract inventors seeking to rival nature.

# 5

## Legs

*...twenty tripods beat*
*To set for stools about the sides of his well-builded hall,*
*To whose feet little wheels of gold he put, to go withal,*
*And enter his rich dining-room, alone, their motion free,*
*And back again go out alone, miraculous to see.*

*Iliad*, Book 18

There are many similarities between human arms and legs. Robot legs have passed through an evolutionary design track similar to that of arms. Robotic legs started as upside down cranes that lacked dexterity in the hips and ankles, just as arms lacked dextrous shoulders and wrist. Multiple SCARA-type arms have been applied as legs, and so have pogo stick designs, all in the hope of finding a simple mechanical solution. But like the arm, the human leg has seven degrees-of-freedom. The hip has pitch, yaw, and roll, although to a lesser degree than the shoulder. The knee is comparable to the elbow, as it provides extension and retraction of the foot. Ankles, like the wrist, have pitch, yaw, and roll but to a lesser degree. Because of the greater loads they carry the hip and ankle joints are by necessity simpler and more rugged. Toes provide fine positioning for balance and traction. Because legs provide primary locomotion and a mobile platform for the arms, they do not require the range of dexterity that the shoulder and wrist do.

Leg technology has grown in tandem with perceived applications. The interdependent design/application relationship continues to produce new designs, as what is accomplished then changes the perception of real world applications. The nuclear industry, with its chronic waste management problem, has led in actual application with the Odetics Functionoid. A more exotic application, in which money and high technology coexist, is space exploration. With one-half of the earth's surface unnavigable by wheeled vehicles, the military has recognized the need for walking machines to traverse rough terrain. The Adaptive Suspension Vehicle (ASV) is the most recent effort. While simplified walking machines for mining have existed for some time, more dextrous legs must be developed as easily accessible ore deposits become depleted and mining is forced into rougher terrain. Forestry has a need for walking machines to harvest trees on steep slopes. Biped robotic research is taking its first steps towards the future.

Many designers have developed walking machines based on reality or dreams. This has lead to designs developed from insects and even pogo sticks. Arthropods such as crabs or spiders, are well suited for their functions and environment, but does this mean that their design is suited as a basis for robot legs? Perhaps the reason for experimentation with nonhuman forms lies in the lack of basic human-like component technology available to designers. Another reason is the difficulty in developing control and feedback systems needed to achieve the balance, that allows a biped to walk upright. Statically stable, quadruped and hexapod robots have been developed to compensate for the present lack of control system sophistication. To escape these limitations designs have been developed that emulate insects because they require easily conceptualized joint technology and simplified control. It is man, however, that has evolved into an all around, general purpose creature adaptable to many tasks and environments. The emulation of man's form in robotics will produce more versatile and practical machines capable of functioning in the environments in which man functions and of replacing man in hostile environments that exceed human tolerance.

## The Human Leg

The human leg begins with the femur connecting to the hip through a ball-and-socket joint (Figure 5.l). Extension and retraction of the leg is performed by the knee, like the elbow functions for the arm. The fibula and tibula connection comprises the second link. As in the lower arm, they may generate limited roll motion by crossing each other. Two axes motion is produced by the ankle, providing the flexibility needed for stability. The structure of the foot resembles the hand. Although less dextrous, the bottom of the foot is analogous to the palm, and the toes to the fingers. Toes provide fine tuning for balance and traction. Humans have lost the need to utilize toes for grasping as they have come to rely on the hand. Interestingly, the fact that some people who have lost the use of their hands have trained themselves to handle articles and even paint by manipulating brushes with their feet is indicative of the fine control that lies dormant in the human foot.

Musculature of the human leg follows the same design rules of the arm. The muscles that actuate the ankle are located in the calf for compactness and to minimize inertia. This concept is carried on up the leg to the hip and buttocks.

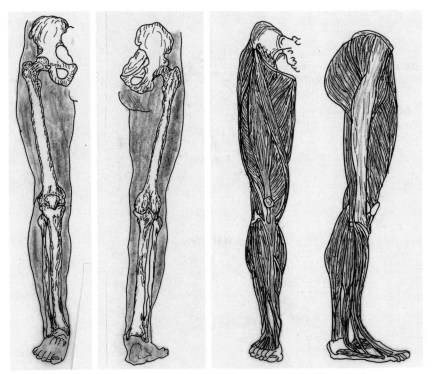

*Figure 5.1 Bone and Muscle Structure of the Human Leg*

Flexion/extension (pitch) of the leg femur is 120 degrees forward and 30 degrees hyperextension for a total of 150 degrees of travel (Figures 5.2 and 5.3). Abduction/adduction (yaw) is 100 degrees. Inward and outward rotation by the hip is 85 degrees, although hyperextension up to 180 degrees is possible. This pitch, yaw and roll motion provides the human leg with its great flexibility. The knee is capable of 90 degrees flexion/extension (pitch) with an additional 45 degrees of hyperextension. Approximately 30 degrees of roll may be obtained from the knee by virtue of the fact than its double ball-and-socket condylar structure accommodates movement other than just flexion and extension under certain conditions.

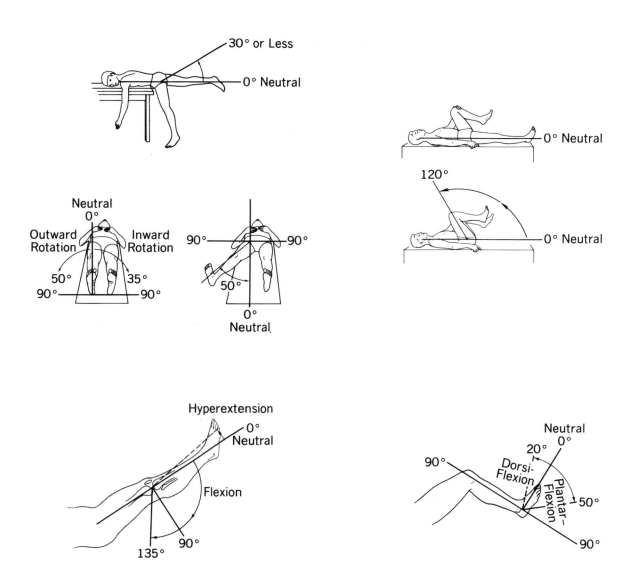

Figure 5.2  Kinesiology of the Human Leg

When the leg has been flexed at the knee, the collateral ligaments become slackened, making it possible to rotate the leg on the thigh, though in a no load condition, for a total range of about 50 degrees. This is analogous to the roll motion of the forearm, although achieved by different means. One example is turning while walking; the foot needs to be repositioned relative to the body for a new direction to be achieved. The ankle possesses a total of 70 degrees of pitch (dorsiflexion/plantarflexion) like the wrist. Yaw (inversion/eversion) is 35 degrees, and has a range of motion bias similar to the wrist (Luttgens and Wells 1982).

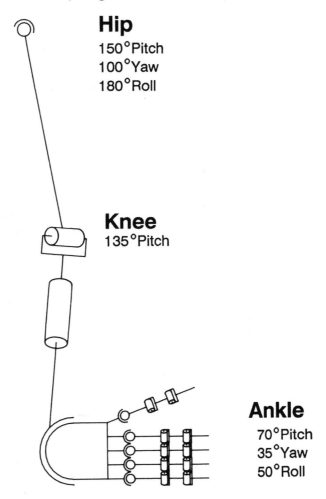

*Figure 5.3 Kinematic Model of the Human Leg*

# Types of Robot Legs

Simple legs for multileg systems are the most common concept. Emulating insects, these arthropod machines typically have six legs. Basically a double tripod, such a structure is inherently stable as one tripod is always in contact with the ground. Limitations in component and control technology have required multileg systems, as the means are only now being developed to achieve bipedal stability. The next generation will see more sophisticated anthrobotic designs utilizing more dextrous and thus more capable multiaxis joints as in the case of arms, wrists and hands.

As with robot arm technology, using the human leg as a model can aid in generating advanced robot leg designs. Transference of robot arm technology to legs is also useful; as stated above, the human leg is a simplification of the human arm, and the same theory may be applied to derive robot legs from robot arms. The need for a general solution to robotic legs, one that is versatile for both two, four or six legged vehicles is clear. Reinventing the wheel has seemed to be the sole solution to the robot leg dilemma. A legged vehicle for autonomous weapon and material carrying vehicles that can overcome the uneven terrain encountered in shell-cratered battlefields, conventional tank traps, or rocky and mountainous regions has been identified for 20 years. However, the ability to meet this need via legged vehicles has been hampered by the lack of adequate joint mechanisms and controls. Joint technology is a key problem in the development of such vehicles, because hip and ankle joints require at a minimum pitch and yaw motion about a common center with remote location of drive sources analogous to our muscles and joints. Transference of wrist joint technology will accomplish this. The lack of simple, compact, cost effective, and reliable actuator packages has also been a major stumbling block in current designs. Ineffective joint designs lead to unwieldy vehicles that compensate for the instability of their simple joints by means of additional legs. Abandonment of these obvious solutions seems only to come after costly and time consuming trial and error. Although the developers are themselves equipped with a pair of superbly designed legs they refuse to follow any of the design lessons these embody.

Advancing to more complex joint morphology does create several challenges. A two legged or four legged vehicle using animal equivalent joints also requires an animal equivalent control system (Raibert 1991). This would require sensors in the joints and actuators as well as controllers that coordinate the motion to varying terrain. The incorporation of vision, ultrasonic sensing, and other basic animal senses would be vital (Klein 1987).

The robot knee is going through evolutionary growth stages similar to those of the elbow. Telescoping knees are attractive from the standpoint of low cost and simplicity, but they limit the ability of angular positioning of the foot. Simple hinge-like joints have given way to parallelogram linkages as seen in the Odex-1 and Adaptive Suspension Vehicle. Perhaps the future will see knees with shock resistant hydrostatic bearings and integral deceleration limits as in the human knee.

## Gaits

A interdependence exists between the mechanical design of legs and gaits that is fundamental to performance control. The sophistication in control must be matched by sophistication in mechanical design in order to produce a useful machine. Gait analysis can provide data on stability, performance on curved or uneven terrain, and dynamic interaction of components and power efficiency. Although an understanding of animal joints and gaits (the corporate motion of legs) is now generally understood, researchers have chosen to develop low cost, simple concepts that can be easily controlled. Attempts to bypass the control issue, as in the case of the Walking Truck, through teleoperation have not been practical.

Fundamental to control of legged vehicles is a understanding of gaits: "A gait of an articulated living creature or a walking machine is the corporate motion of the legs, which can be defined as the time and location of the placing and lifting of each foot, coordinated with the motion of the body in its six degrees of freedom, in order to move the body from one place to another." (Song and Waldron 1989) Animals select various gaits depending on terrain and desired speed. Sensory data is required to select intelligently the optimum mode for the desired operational speed or terrain. Knowledge of gaits provides the designer with a base of which control algorithms for walking machines can be written.

Two forms of gait analysis have been developed, analytical and graphical methods. Of the two forms, graphical is the most suited for discussion in this book as it provides a quick understanding of the geometrical representation of the gait and is the most useful to mechanical designers. The earliest form used was motion pictures. Muybridge (1899) used batteries of still cameras to record animal and human motion. Footfall formulas are another early modern technique for representing gaits. The animal is observed from above as it transverses from left to right. Foot contact with the ground is indicated by the black circles with the arrows indicating direction (Figure 5.4). A more elaborate technique was developed by Hildebrand. Known as the "gait diagram," this not only records the sequence, placement, and lifting of different feet, but also provides support and transfer phases of each leg as a function of time.

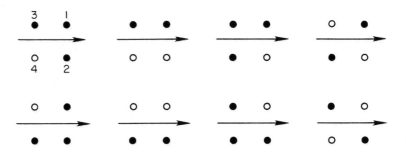

*Figure 5.4  Footfall Diagram for Four Legged Animal*

The event sequence diagram is used for periodic gaits because it provides information about placing and lifting of the leg. Read clockwise, the circumference of the circle represents the cycle time (generally normalized to 1).

Other forms of gait diagrams include the successive gait pattern, which shows the successive support patterns of a gait, with the arrow showing direction (Figure 5.5). The stationary gait pattern incorporates support patterns at different times and the corresponding body motion for each support pattern. It is useful for studying the gait stability margin of periodic gaits. The lateral motion sequence presents the machine in profile in a multiple sequence and provides center of gravity as an aid to understanding obstacle crossings. All of the above may be computer generated and are useful for building up more complex two and three-dimensional graphics that provide walker stability and the gait adjustment necessary to traverse obstacles. Most realistic of all methods, it may simulate various gaits and their performance over unique terrains and obstacles.

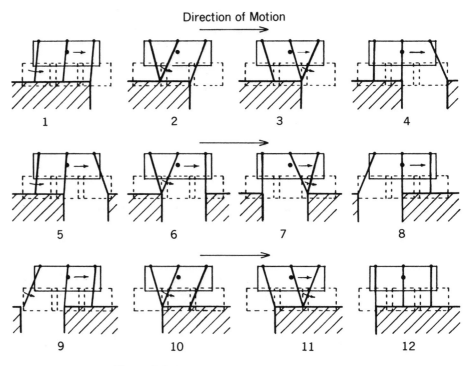

*Figure 5.5 Successive Gait Pattern Diagram*

## New Walkers

Interestingly the two examples of new walkers discussed here are of recent vintage but are placed here because they represent simpler non-anthropomorphic design. One wonders why they were not conceived of initially. Perhaps because these practical designs lacked the romance of their anthropomorphic counterparts. Nonanthropomorphic, these designs are analogous to gantry and SCARA robots. Walking machines that have legs ganged together are collectively grouped as frame walkers. Several designs have been patented (Kroczynski 1987) and at least one implemented for undersea operations (Martin and Kuban 1985). The Walking Beam Machine was developed by Wendell Chun of Martin Marietta Astronautics in 1989. The Walking Beam's purpose is the NASA Mars Rover Sample Return mission that anticipates a 1998 Mars mission (Figure 5.6) (Pramberger 1990). It is designed to negotiate a variety of rough terrains, as hazards on the Red Planet include soft, loose soil and jagged lava flows (Chun 1992). The Beam's specifications include providing a stable platform for drilling core samples, taking soil samples, and performing analysis. Upon accomplishment of its tasks the robot returns carrying its payload of precision cargo. Specifications require: (1) Crossing a 1m ditch, (2) negotiating a 1m step, (3) climbing a 60 percent grade with a hard surface, and (4) climbing a 35 percent grade with a loose, sandy surface.

The Walking Beam is a step towards simplicity over previous designs. The low cost, quickly assembled, 1/4 scale working model (Figure 5.7) was a clear indication of future success . Two nested platforms, one within the other, that alternately translate with respect to each other, propel the Beam. Imagine two nested tables: an inner one with four sliding legs and an outer with three legs (Figure 5.8). The legs of both "tables" are capable of telescoping vertically (Figure 5.9). The translating horizontal motion is decoupled from the vertical motion of each leg. Efficiency and simplicity of control is greatly increased, eliminating power dissipation from feet digging into and out of loose soil like the Hexapod's. Size is 19.7' (6 m) by 19.7' (6 m) by 6.57' (2 m) tall. The large size provides maximum stability over uneven or unstable terrain.

Figure 5.10 depicts the Beam translating and turning. The turning sequence is as follows: lifting and rotation of the inner platform relative to the outer "T" frame, followed by raising of outer frame or vice versa. Thus the outer T frame can be rotated in a new direction when lifted by the inner frame.

The 1/4 scale working model was built and successfully tested on a budget amazingly low in walking machine technology (Chun 1989). Constructed of aluminum except for the footpads, the Beam's weight is 361 Lbs (163 Kg), including electronics. The electronics are housed in the payload base. A 24V dc power source allows battery operation, although the unit is currently tethered and powered through a portable generator. Sensor equipment includes potentiometers on the motors and load cells that sense the forces on the footpad; in addition, microswitches sense foot contact with the ground. Microswitches and electromagnetic brakes provide fail-safes for all actuators.

*Figure 5.6  The Walking Beam by Martin Marietta*

*Figure 5.7 Working One-quarter Scale Walking Beam Model*

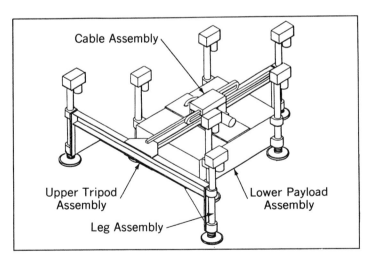

*Figure 5.8 Layout of Beam*

Cable Assembly

Upper Tripod
Assembly

Leg Assembly

Lower Payload
Assembly

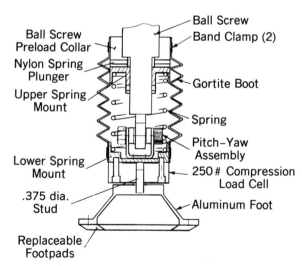

*Figure 5.9 Leg/Foot Detail*

Ball Screw
Preload Collar

Nylon Spring
Plunger

Upper Spring
Mount

Lower Spring
Mount

.375 dia.
Stud

Replaceable
Footpads

Ball Screw

Band Clamp (2)

Gortite Boot

Spring

Pitch-Yaw
Assembly

250 # Compression
Load Cell

Aluminum Foot

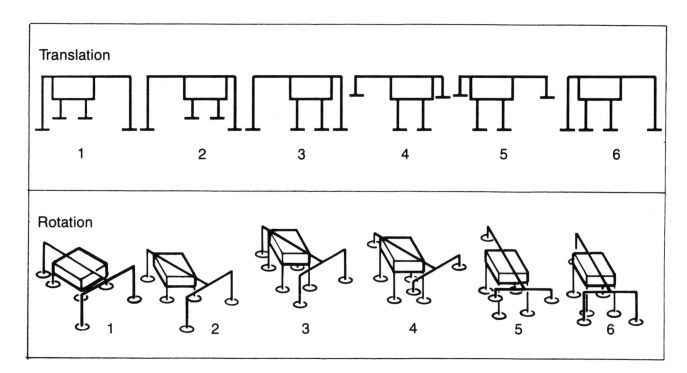

*Figure 5.10  Walking Beam Translation and Turning*

## Carnegie Mellon Walker

Described as an Autonomous Mobile Exploration Robot (Figure 5.11), it is similar in principle to the Walking Beam. Built at Carnegie Mellon University, the aluminium Ambler weighs in at two tons and stands over l9' (5.8 m) tall at full extension. It is the largest and heaviest walking machine built to date. The Ambler uses six telescoping legs divided into sets of three pairs that work in opposition. The legs are moved by articulated SCARA-type arms. The body is propelled in a motion similar to cross-country skiing. With six legs, it substitutes SCARA-type arms for the traversing slider arrangement of the Walking Beam. The legs vertically extend and retract as in the Walking Beam. Like the Walking Beam, Mars is the planned destination of this machine (Anon. 1990). More complicated than the Walking Beam, it does eliminate the roller sliding elements. The action of Ambler is described by the university as follows.

*Figure 5.11  Carnegie Mellon Walker (Ambler)*

A single leg reaches out in front of the other, places itself firmly on the ground like a ski pole, and then pulls the machine forward. It may traverse a 5 foot (1.5 m) ditch. In the event of an accident or failure any functional leg can reposition itself to substitute for any failed leg. Thus, the Ambler retains its ability to walk even if it loses mobility in two legs. Its two stacks are connected to an arched body structure that involves enclosures for power generation, electronics, computers and scientific equipment. Because the drive motors that support the Ambler's body are separate from those that propel it the robot remains level whether it walks on flat or rugged ground. The design provides a stable platform for sensors, scientific instrument, and sample acquisition tool (Locke 1990).

The Ambler's vision system creates a 3-D map of the surrounding topography in relation to sampling targets. It plans the best route to the target and the Task Control Architecture determines the optimum number of steps (Watzman 1990).

## General Electric Walking Truck

This 3000 Lbs (1363 Kg) machine is the first true walking machine and has inspired several imitations (Figure 5.12 through 5.15). Applications foreseen included loading bombs into overhead bomb bays and negotiating fast moving streams and river banks (Mosher 1967). Developed by robotics pioneer Ralph Mosher for General Electric in the 1960s, it was controlled by a human operator and was capable of loading a 500 Lbs (227 Kg) payload at speeds up to five mph (Liston 1968). An external 90 hp gas engine provided power. The legs contribute 6.7' (2 m) to the overall height of nearly 11' with a length of over 11.7'. Each foot exerts a nominal 1500 Lbs of force at 5 mph. Force reflection of 1 part in 120 is reflected back to the operator. Large mirrors were added as an essential aid to controlling the hind legs. Hydraulic vane actuators and cylinders were used in the hips and knees. Tested outdoors using a roll cage to protect the operator, the machine worked, but the primitive analog/human operator control technology required constant input from the operator. The analog control was described as follows: "Force reflecting servos necessary to move the machine are powered by a special motor and pump system. The rear legs respond to an operator leg action; the forelegs respond to the motion of the operator's arms." The instability of the joints, plus the simple controls led to a quadrupedal machine that required the operator adapt to it, instead of the other way around. This resulted in the obscenity of a "labor saving device" that forced a person to regress to thinking on all fours. The operator simultaneously controlled twelve position servos and was required to respond readily to 12 inputs. This led to a unacceptable level of operator fatigue, which relegated the hardware to the warehouse.

*Figure 5.12  General Electric Walking Truck*

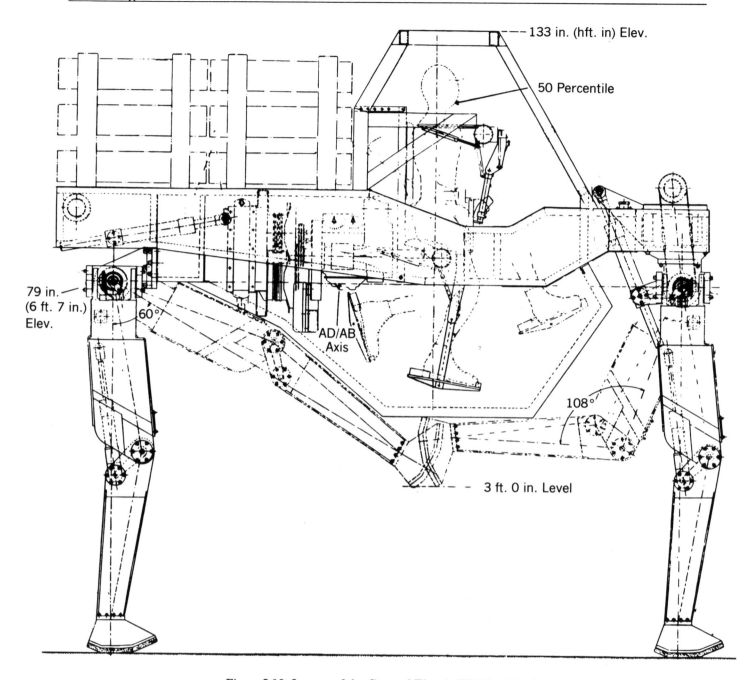

*Figure 5.13  Layout of the General Electric Walking Truck*

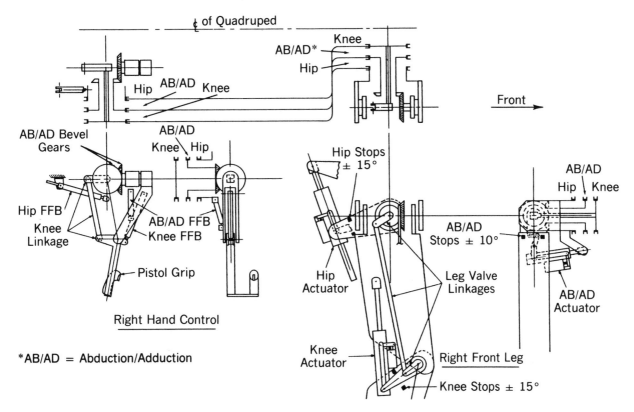

*Figure 5.14 Schematic Diagram of Right Hand Control*

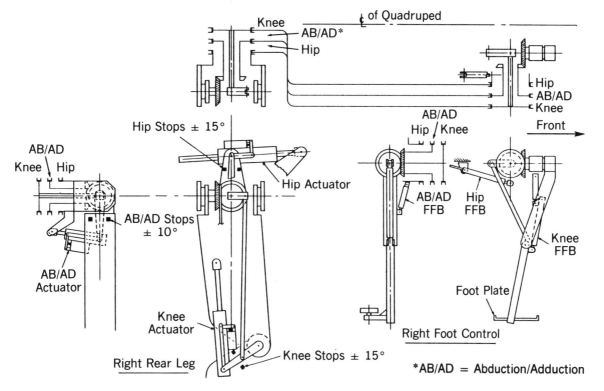

*Figure 5.15 Diagram of Right Foot Control*

## Phony Pony

The Phony Pony (Figure 5.16), also known as the Walking Horse and California Horse, was based on the General Electric Walking Truck. The Phony Pony was a pioneering effort to use a computer to control a legged vehicle. This is an obvious advance over the mechanical analog techniques previously used on the Walking Truck. Built in 1966 at Ohio State University by R.B. McGhee and A.A. Frank, it had four legs and weighed 110 Lbs (50 Kg). Speed was .5 miles per hour. The legs consisted of a one degree-of-freedom hip and a one degree-of-freedom knee for a total of eight independent degrees-of-freedom. Power was supplied by two 12 Volt automobile batteries through a trailing umbilical. Cannibalized handheld electric drill motors driving worm gears were used for actuators. Limit sensors provided for protection from hyperextension. The Phony Pony is capable of forward and backward motion, as well as locking its joints. A passive suspension system also was included for permitting vertical excursion of each leg relative to the vehicle body. Two modes of walking were achieved: a quadruped crawl and the diagonal gait, or trot, but it was not capable of turning. In the trot, broad feet are required for stability. Due to the lack of true pitch and yaw motion of the hips and to a lesser extent of the ankles, six legs were required for stability. Developed prior to the microprocessor era, computer control was via an electronic sequencer made of flip-flops. Utilizing finite state control theory, the controller was a six or seven state machine. Each state activated the legs in various stages for a particular gait. Each leg had its own controller phased to the other legs through interlocked signals (Todd 1985). Phony Pony walked with a pronounced stiff legged gate. Clearly robotic, it was not until the 1980s that computers became capable of controlling legs with fluid motion (McGhee 1976).

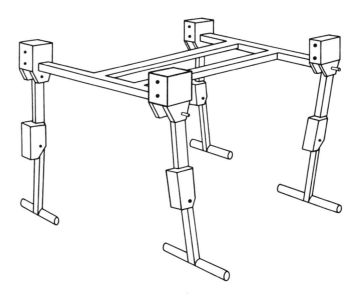

*Figure 5.16 Ohio State University Phony Pony*

## Ohio State University Hexapod

Taking its first steps in 1977, the Hexapod is 51" (1.3 m) long, 55" (1.4 m) wide, and weighs 220 Lbs (100 Kg) (Figure 5.17). Vehicle speed is a few feet per second (Todd 1985). The lack of knees and ankles makes it insect-like in appearance. Initially funded by a $405,000 National Science Foundation Grant, in 1978 it received Senator William Proxmire's "Golden Fleece" award for wasted government money (Omni 1988). The perception was that hundreds of thousands of dollars for a laboratory demonstration device that had no immediate practical application was a poor investment. However, one can make the same argument about the size of congressional salaries. In any case, if politicians had the vision to do basic research they would be scientists, not politicians. In spite of the senator's opinion, NSF funding increased and was joined by additional DARPA money. Not to be outdone in parsimony, the designers used proven low cost electric drill motors to drive worm gears that not only boosted torque but provided self-braking (Figure 5.18). The Hexapod is an example of scientific thrift that belies the senator's charges of financial waste.

Pitch and yaw hips with a nonhuman bias produce joint torque of 200 ft-Lbs. Various gaits, obstacles and low stairs have been negotiated. Power is supplied though an umbilical at 115Vac 60 Hz using half-wave triac (a type of thyristor) phase control. Triac trigger signals are produced by a PDP-11/70 computer through an umbilical which also provides transmission of joint position sensor data, motor speed, leg force, and body

*Figure 5.17 Ohio State University Hexapod*

attitude. Gyroscopes, proximity sensors, and a camera system have been added since its inception (McGhee 1984).

Some of its accomplishments include walking with different gaits, climbing shallow stairs and obstacles, turning, active compliance, use of sensors and microprocessor leg coordination, analysis and control. In the history of this one machine we can see a parallel to robot arm controllers, which began as simple sequencer devices but evolved with modern microprocessor control. As a consequence of the rather simple control technology then available the mechanical technology also had to be kept simple.

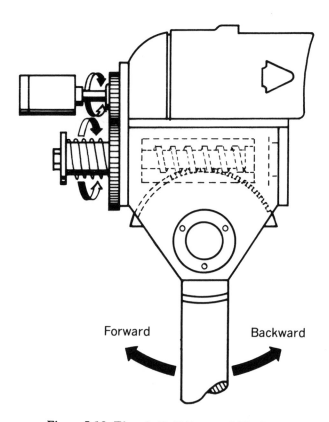

*Figure 5.18 Electric Drill Powered Hip Joint*

## Odetics Functionoid

The Functionoid (Figure 5.19) by Odetics, Inc. of Anaheim, California, has achieved perhaps the greatest fame of any modern robot system and has appeared on television and countless magazine covers, as well as in numerous newspaper articles (Russell 1983). Displayed in 1983, it weighs 370 Lbs (167.8 Kg) and also has six legs. A 24V, 25 A/hr aircraft battery provides a self-contained power source with up to one hour of operation. When walking 450 Watts of power are consumed, but at rest a mere two Watts are used. When moving 1 inch/second the Odex-1 model of the Functionoid uses 100 Watts of power and approximately 150 Watts at double this speed; thus, a substantial power drain occurs at 10 inches/second. On the reverse stroke the motors act as generators, storing the surplus power in the battery. Brakes on the motors also help to conserve power. A double tripod, the system raises three articulated legs at a time to advance. The other three support the machine. The legs are

*Figure 5.19 Odetics Functionoid*

a parallelogram design similar to the one used in industrial robot arms. Ball screws operate the hip pitch and ankle pitch motion, and hip yaw swings the assembly via a gear drive for side-to-side motion (Figure 5.20). This simple and effective system differs significantly from the ASV described below. In the tucked profile, with the legs folded, the Functionoid has its slimmest profile with a width of 27" and a height of 48". The tall profile raises the payload area to 78" when the legs reach an maximum extension of 65". Squatting, the profile spreads the legs to 72" apart with the payload area only 36" off the ground. The Functionoid for maximum flexibility has 105" between legs in the low profile. It is capable of climbing steps 22" high. Eighteen motors are employed, the largest motors being used for lifting. Odex-2 features tactile and kinesthetic foot sensing. Odex-3 replaces the pantograph legs with cable driven telescoping ones. Built for the Electric Power Research Institute (EPRI), it is capable of raising itself several meters to gain access to restricted areas in nuclear power plants. Thus, the Odex-3 legs extension/retraction is accomplished in a different manner, analogous to the telescoping elbow of the early Unimate robots.

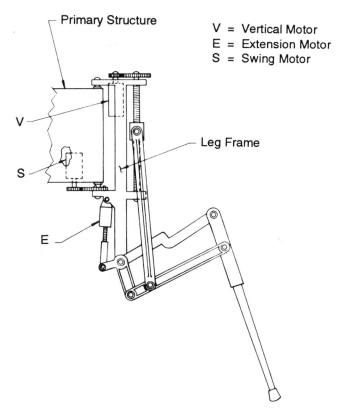

*Figure 5.20  Odetics Functionoid Pantograph Leg*

## Adaptive Suspension Vehicle

This multimillion dollar, DARPA funded project is an excellent example of robotic "Big Science": very interesting, but a person wonders if more could have been done to produce a practical machine (Figures 5.21 through 5.23). Designed at Ohio State University as an alternative to tracked vehicles, such as tanks, the Adaptive Suspension Vehicle (ASV) may be used outdoors. Its dimensions are 16.4' (5 m) long, 12.8' (3.9 m) high, and 5.2' (1.6 m) wide. It weighs approximately 5720 Lbs (2600 Kg). Cruise speed is 2.25 m/s, while top speed is 3.6 m/s. A 0.9 liter (900 cc), 70 KW motor cycle engine drives a 0.25 KW/hr flywheel that transfers power to a group of 18 variable displacement pumps interconnected via parallel line shafts that run the vehicle's length through toothed belts. The actuators are double acting hydraulic cylinders powered by the pumps. Loading of each leg is obtained by reading the hydraulic pressures across the

*Figure 5.21  Adaptive Suspension Vehicle (ASV)*

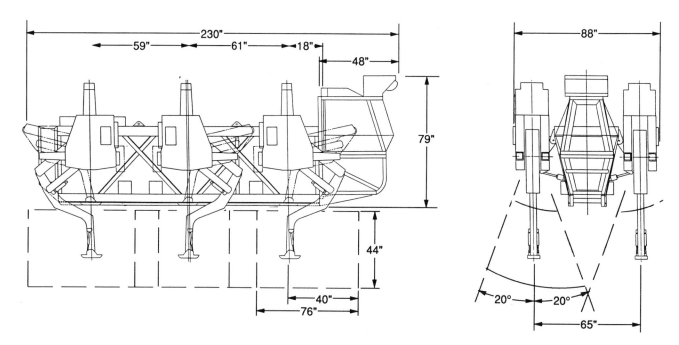

*Figure 5.22  Dimension of the ASV*

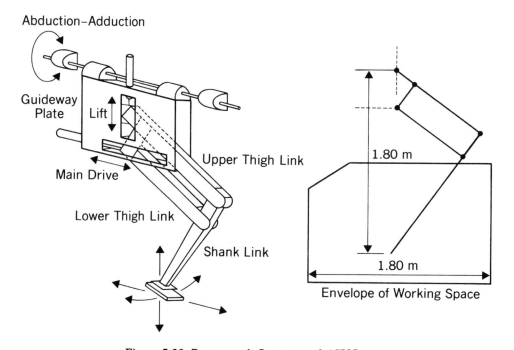

*Figure 5.23  Pantograph Structure of ASV Leg*

multiple actuators (Waldron 1984). It requires a driver, like the General Electric Walking Truck, albeit with much less burden for the operator due to advanced computer control (McGhee & Waldron 1985). The operator uses a six axes joystick for general motion of the ASV with the control of the individual limbs and the coordination of the legs controlled by computer. Sophisticated optical radar mounted on the top of the cab provides terrain information that is used for foothold selection and body orientation in some control modes. A vertical gyro provides a vertical reference, while rate gyroscopes and accelerometers provide information on body movements. All six legs are fitted with three axes force sensors to determine the three force components at the ankle. Five control modes may be selected by the operator: 1) Utility mode—checks out all major vehicle subsystems and allows reprogramming of the control computer. 2) Precision footing—the most manual of all modes; the operator selects the body or an individual leg as the control subject. Body motion with six degrees-of-freedom or foot motion with three degrees-of-freedom may be selected. 3) Close maneuvering mode—A general purpose mode for use in easy terrain; patterned gaits are used. In this fastest of all modes, data gathered from the scanning rangefinder is used to adjust body attitude in anticipation of gradient changes, but not footholds. 4) Terrain following—the most automated of all modes, it is intended for use in moderate to severe obstacle fields. Footholds are selected from the scanning rangefinder data and a free gait algorithm is used. 5) Follow the leader mode—used when crossing large obstacles, the scanning rangefinder analyzes the geometry of the obstacle and feeds information into onboard computers, decreasing significantly the operator control burden.

A pantograph structure is used on the legs (Figure 5.23). Extension and retraction is powered by linear motion of the actuator. The upper portion provides lift, and abduction/adduction is generated by pivoting the entire mechanism. Prismatic, sliding joints using rollers and tracks comprise the elements in this complex design. They are used throughout but are difficult to enclose from the outdoor environment, and because of this the design is less cost effective than other similar joints. The hip achieves pitch and yaw in a very cumbersome way, that incorporates the knee into the hip mechanism with no ankle actuation.

### Omni-Leg

Proposed by Ross-Hime Designs in 1987, this is an outgrowth of my early wrist design, the Hydraulic Servo Mechanism (Rosheim 1980). This is my attempt at producing a practical design that distills the basic motions of the human leg with as few parts as practical. Simple, rugged actuator packaging is a valuable feature of the joint (Figure 5.24). I have taken my design cue from nature and based it on the ball-and-socket type joints in the human hip and ankle. All moving parts are enclosed in a solid metal housing, and the joint is sealed dynamically by an O-ring in the housing lip. Counterbalancing would be required for multilegged walking machines constructed from the joints. Counterbalancing the weight of the walking machine employing articulated legs is easily accomplished by placing preloaded springs over the hydraulic cylinder rods. Roll rotation, which is common to animal hip joints, may be added by connecting a rotary vane actuator to the joint output drive shaft. The hollow design of the joint allows wiring or hoses for sensors or additional actuators, such as a roll actuator, to pass though the joint unencumbered.

Transferable technology is adopted in the design of this joint. Hydrostatic suspension of ball-and-socket joints applies design technology similar to that used in rocket nozzles, as shown in Figure 5.25. Self-lubrication is accomplished by means of continuous injection and recirculation of hydraulic fluid under pressure between the ball-and-socket surfaces and trunnions. Hydrostatic suspension assumes the function of synovial fluid in animal joints. The result of this emulation is a design requiring little preventive maintenance. As mechanical sophistication and mobility of robots approaches human or other biological models, the robot is likely to fall or stumble like its human counterpart. For this reason, a machine that is rugged and capable of recovering from minor accidents is both functionally and logistically necessary. Figure 5.26 shows Omni-Legs used to create a walking truck.

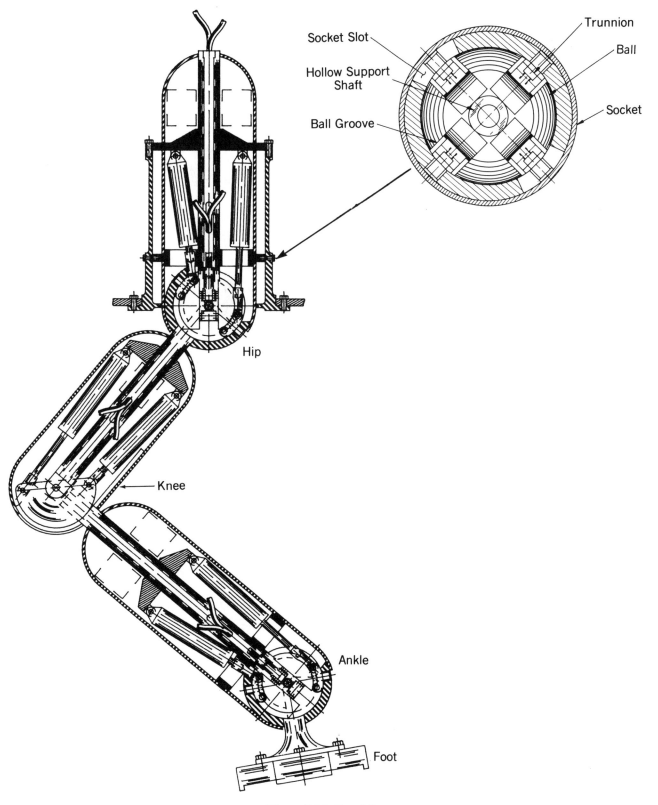

*Figure 5.24 Omni-Leg*

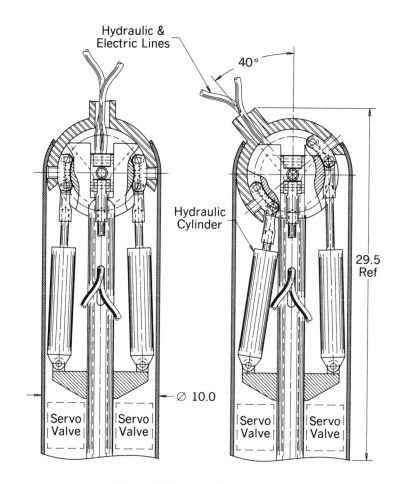

Figure 5.25  Omni-Leg Drivetrain

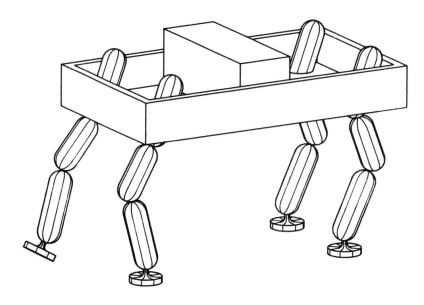

Figure 5.26  Omni-Legs Used to Create Walking Truck

## Bipeds

The nature of biped locomotion is controlled instability, which some have likened to controlled falling. Bipedal locomotion requires a great deal of sensor information. The soles, when in contact with the floor, are analogous to an arm's endeffector and the floor is the workpiece. The more complex the surface, the more dextrous the legs must be.

Bipeds are evolving through stages similar to robot arms. The initial reaction of the designer to the problem of bipedal locomotion is that it may be simplified with fewer degrees-of-freedom or less range of motion. Whereas limiting range of motion may be valid until running robots are required, dexterity is not. Dexterity is necessary for stability and must be a design objective for the hip and ankle.

Perhaps the earliest successful biped was Steam Man by George Moore in 1893 (Figure 5.27). He was powered by a 0.5 hp gas fired boiler and reached a speed of 9 mph (14 kph). Stability was aided by a swing arm that guided him in circles . Traction was aided by heel spurs, smoke flowed from his head, steam from the nose and a pressure gauge was conveniently mounted in his neck (Ord-Hume 1973).

*Figure 5.27  Steam Man*

## Wabot

The Japanese have been leaders in two legged walking robot research. Advantages to two legged vehicles include narrower width and consequently greater maneuverability in confined spaces. WL-5, an early two legged walker developed at Waseda University (Figure 5.28), took its first tentative steps in 1972. Hydraulically powered, it uses disproportionately large feet for stability. A shuffler more than a walker, WL-5 achieved "static walking." Invented by Ichiro Kato at Waseda University, Tokyo,

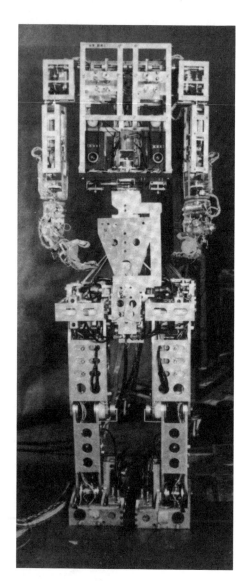

*Figure 5.28  WL-5 Built at Weseda University*

Japan, the design displays the first two evolutionary stages: (1) the over-simplified design deletes the knee and ankle joints; (2) distributed actuators with standard motors or hydraulic actuators arrayed for powering the for hip, knee, and ankle joints. The latter stage is embodied in the WL-10RD which started walking in 1984 (Figure 5.29). It has six degrees-of-freedom per leg and walks forward, backward, and sideways, but can't turn. Not yet ready for the Olympics, it takes 4.8 seconds for a stride of 17.7" (45 cm) and weighs 176 Lbs (80 Kg), with a maximum step rate of 1.3 seconds per step. Its inventors describe the walking motion as taking place in two phases: (1) *Change over phase* (leg support-exchange phase), in which only the ankle joints of the legs move and the center of gravity is transferred from the rear leg to the front leg by moving the ankles only. In this phase, the robot, as an inverted pendulum, leans forward. This phase causes a large shock, which is absorbed by providing the ankles with flexibility. (2) *Single support phase*, in which the center of gravity remains almost unmoved, and the swing leg is moved from the backward position to the forward position. In this phase, a reference signal is given to each joint such that the zero moment point (ZMP) moves in the rectangle formed by the four pins in the sole of the supporting leg (Vukobratiovic 1975)." The WL-11 (Waseda Hitachi Leg) exhibited at Expo 85 has an onboard computer and hydraulic power supply with a static walking capacity of 13 leisurely seconds per step. WL-12, the latest model in the series, succeeded the earlier models in 1987.

*Figure 5.29  WL-10RD*

## BLR-G2

Developed at Gifu University by J. Furusho and A. Sano, BLR-G2 followed in the footsteps of Wabot. BLR-G2 builds on the Wabot design and adds sophisticated, motorized ankles, that increase its agility (Figures 5.30 and 5.31). However, it suffers from using stock components that are detrimental to performance, although they may be cost and time expedient. Speed is .18 m/s with a stride length of 25 cm and a stride period of 1.4 seconds (Furusho and Sano 1990).

A bipedal walking robot is required to support and balance its whole weight on one sole. This, plus rigidity and strength, puts high demands on the sole. Pitch is provided by the belt drive and roll via a flexible shaft that drives a screw/nut actuator.

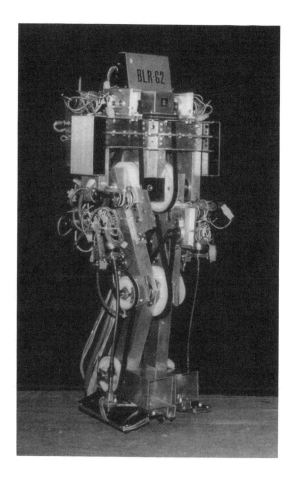

*Figure 5.30 BLR-G2 Developed at Gifu University*

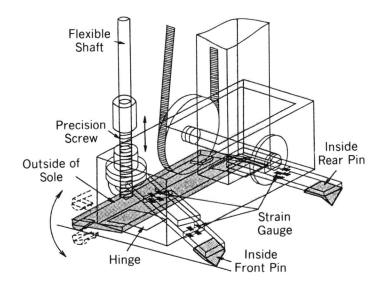

*Figure 5.31 Foot Mechanism of BLR-G2*

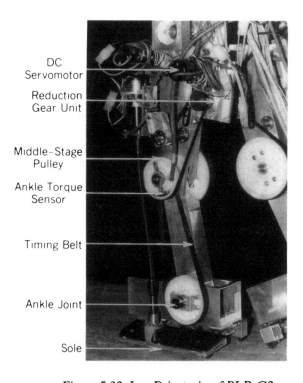

*Figure 5.32 Leg Drivetrain of BLR-G2*

The sole senses the reaction with the floor and thereby realizes flexible movement through force control (Figure 5.31). In the foot, three sets of strain gauges are used as pressure and ankle torque sensors. The strain gauges near the screw/nut actuators produce a force signal used to control the floor reaction above the sole. In the ankle torque sensor, the ankle gearhead motor is mounted near the hip. The middle stage pulley shaft also serves as the knee joint shaft. The ankle torque sensor is mounted in a midstage pulley to create a strong and compact ankle.

Inclinometer and angular rate sensors measure pitch and roll angles as well as angular velocities about the pitch and roll axes. An ultrasonic speed sensor measures forward speed through the Doppler shift. Accelerometers measure forward acceleration of the body at intervals of 20 milliseconds. Body speed for feedback control is extrapolated from the accelerometer and inclinometer outputs.

Hips, knees, and ankles power pitch shafts (Figure 5.32). Soles have a roll shaft. Each leg has four degrees of freedom. Weight is 55 Lbs (25 Kg), and the unit is 38.7" (.97 m) tall. Each articular joint can be driven independently by a combination of a DC servo motor with gearhead and timing belt or flexible shaft. The remote location of drives allows larger motors to power the joints and improves weight distribution, similar to that of the human body. The developers also found that a minimum amount of inertia around the hip joint improves leg control.

Two 16-bit microcomputers were used to control the BLR-G2 hierarchical processing system: one for high-level processing and the other for lower level processing. The time required for information exchange was reduced by receiving all information through the lower-level computer and by utilizing a common memory. Figure 5.33 shows the BLR-G2 making one complete stride.

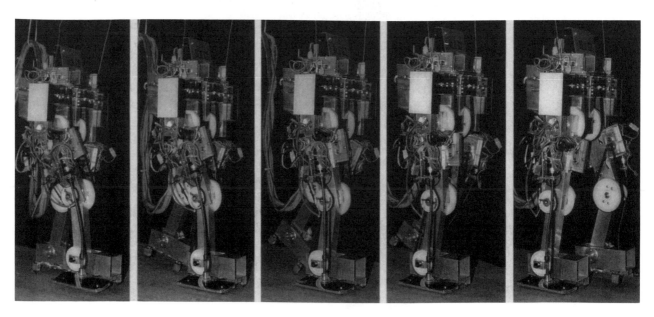

*Figure 5.33 BLR-G2 Taking a Step*

## Dynamics

Most of the machines described in this chapter are statically stable, having at least three legs on the ground at the same time while maintaining their center of mass within the tripod formed by the legs. Statically stable walkers are simpler to analyze, make real-time computations much easier, and are much easier on fragile, laboratory walkers. The standard approach to walking research is to generalize from static machines to dynamic stability later, however only a few people have achieved this.

The dynamics of walking and running have been pioneered through a series of experimental laboratory machines by Marc Raibert at the MIT Artificial Intelligence Laboratory. Raibert insists that dynamic stability was the fundamental issue and he started with a one legged machine to force the issue. The three modes of locomotion are: crawling (insect-like motion that does not require balancing), walking (locomotion that requires balancing), and running (requires balancing and involves period of flight).

Dynamic balancing on one leg has progressed from earlier research at Stanford University in which an inverted pendulum is balanced on a motorized cart. In 1984, the Japanese WL-Lord achieved quasi dynamic balancing. However, it was Raibert that discovered a simple reformation of the dynamic stability problem. He started with a two dimensional (2D) hopping machine, analogous to the locomotion of a kangaroo. Each leg changes its point of support all at once and the leg must be unloaded to do so. The loaded phase is called *stance* and the unloaded phase is called *transfer*. On Raibert's 2D hopper, control is provided for the hip angle and leg thrust, as well as measure hip angle, leg extension, and foot on ground parameters. There are simple control laws for hopping: the desired forward position of the foot is a function of its desired forward velocity and the duration of stance. Control is broken down into three parts: control of hopping height, control of forward speed, and control of body attitude. Body attitude (i.e., hip torque) is a function of position and velocity gains and the desired attitude of the body.

In actuality, the 2D hopper is controlled by a five step state machine. The foot is loaded when it touches the ground. The leg shortens and is in the compressed state. Pressurizing the leg lengthens it and the robot thrusts. The leg achieves full length and is in the unloaded state. When the foot is not touching, it is in flight until it touches down during the landing (loaded) state. At this time the cycle repeats itself. Raibert later extends the pogo stick to three dimensions (3D). The 3D state control machine has the leg going through three phases in one step cycle (Raibert & Hodgins 1992). The three state ares: touchdown (the leg compresses), bottom (the leg thrusts), and lift-off (leg flight). Marc Raibert's work extended later to a two legged walker that resembles two onelegged hoppers supported by a tethered boom. This configuration can perform robotic gymnastics by completing a mid-air somersault while running. Since then, the hopper work has progressed to a four legged (four single legs) runner that addresses other modes of running such as prancing and galloping.

By coordination of the various controllable factors, dynamic balance is achieved. Advantages of dynamic over static systems is the ability to traverse rough terrain (seventy percent of the earth's land surface is inaccessible to wheeled vehicles) and delicate terrain (soft surface that can be torn up). Legged vehicles decouple the path of the body from the path of the feet, while wheeled vehicles have a continuous footprint. As a result, dynamic stability enables a robot to leap over obstacles and to balance in terrain that would destabilize a statically balanced system (Raibert 1989). One must realize that a walking or running machine that is as sophisticated and dynamic as a human or other animal and will share the same faults. Inevitably, it will stumble or fall and need to recover to go on its way.

## The MIT Quadruped

The MIT Quadruped is an example of a dynamically stable running machine (Figure 5.34). Not limited to running, the Quadruped may also trot, pace, bound, pronk and make transitions between these gaits. An aluminum frame supports four opposing legs, hydraulic cylinders, hip actuators, gyroscopes, and computer interface electronics (Figure 5.35). Gimbaled hip actuators provide pitch and yaw and the telescoping legs perform the function of the extend and retraction of the knee. An air spring chamber in the end of each leg provides spring. Linear potentiometers provide sensing for leg length, overall leg length, leg spring length, and contact between the feet and the floor. The gyroscopes provide pitch, yaw, and roll orientation of the body. Hydraulic power is provided through umbilicals.

Control algorithms were written for the various trotting, pacing, and bounding gaits. Two levels of control are used (1) high level: gait control for vertical bouncing motion, stabilizes the forward running speed, and keeps the body level (2) low-level: provides coordination of legs, manipulation of the legs for feet placement, relative forces a pair of legs exert on the ground, and torque the hips experience from the legs (Raibert 1990).

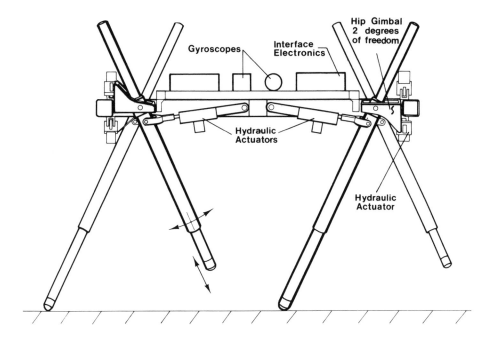

*Figure 5.34  MIT Quadruped*

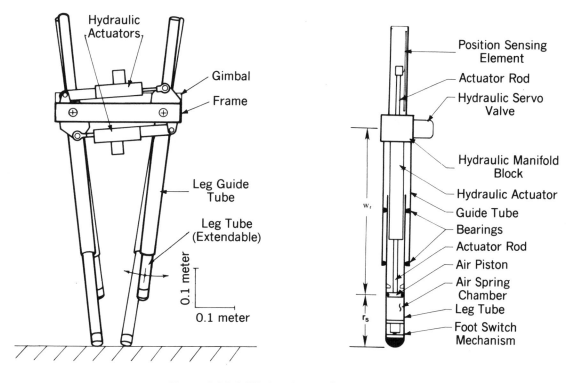

*Figure 5.35  MIT Quadruped Leg Assembly*

## Conclusion

In this chapter we have examined the human leg and identified its structure and three-dimensional motion characteristics. Types of robot legs were also examined and their parallel to robot arm evolution. Efforts to produce low cost, insect-like walkers and consequently insect-like motion were reviewed. A brief look at graphical gait analysis suggests that regardless of the number of legs, a successful walker must be designed that is based on an understanding of the corporate motion of the limbs.

Several well known designs were presented and their merits and demerits examined. The advent of the biped robots in Japan with their obvious appeal, represents an attempt at overcoming the greatest challenge in walker design. Unique among walker technology, it is anthropomorphic in structure and kinematics. Its successful incorporation in a mobile design awaits the development of enhanced control technology.

The whole concept of walking machines is predicated on the ability to transverse rough terrain that wheeled or track vehicles cannot. To make a truly effective walking machine, advances in sensors, computer vision, and artificial intelligence are required. These advances are required to successfully integrate the kinematics, dynamics, spatial reasoning and other factors needed for coordinated foot placement, leg movement and body stability. Development of artificial intelligence techniques sufficient for a practical walking machine is part of the great challenges facing anthrobotics.

The marriage of advanced controls and structural kinematics are critical if mobile robots capable of autonomous movement in three dimensional environments are to be developed. After all, what use is a low dexterity legged robot that stumbles and breaks down constantly? In the walking machines and especially in the bipeds we see one of man's most distinctive features, the ability to walk upright on two legs. With this we come one step closer to the machine that man has sought for centuries: an anthrobot made in the human image.

# 6

# *Anthrobots*

*Teleautomata [Anthrobots] will ultimately be produced, capable of acting as if possessed of their own intelligence, and their advent will create a revolution.*

Nikola Tesla, 1898

## Introduction

The previous chapters have shown the various components—arms, wrists, hands, and legs—which require integration on the minor subsystem level. In this chapter we will examine the major subsystems of the human as a frame of reference for seeing the evolution of robots away from stationary factory devices to fully functional systems that will comprise mobile anthrobots. At present there are few missing links in the evolution to anthrobots. Numerical control technology has developed into versatile controllers for robots. Machine tools and hard automation advances can form the mechanical basis for robots. Computers, once dedicated and single purpose machines, are evolving into general purpose thinking machines.

High-level functioning such as human arm dexterity and mobility makes system integration crucial: the wrist must be matched to the arm and hands, control and artificial intelligence (AI) also must be matched, and so forth. Unless weight and mass distribution are low and mechanical efficiency is high, self-contained power requirements cannot be met. Anthrobotic analogies must be developed for the five human senses:
- Sight
- Hearing
- Smell
- Taste
- Touch

Multisensory input controls and artificial intelligence are crucial to high level functioning in unstructured environments. These technologies, along with a discussion of artificial intelligence, form the basis of this chapter. The chapter concludes with early anthrobots, man-amplifiers, and the design of an anthrobot that mirrors man.

The creation of a mechanical facsimile of man hinges on the functional duplication of the human structure's mechanisms. Examples of the these are the spine, shoulder, wrist, and ankle. It is these three-dimensional space mechanisms that produce the signature "life-like" motion that has been missing for so long in robotics. In the conclusion of each previous chapter I presented man equivalent mechanisms that in a practical manner reproduce human-like motion. They are mechanical equivalents of human actuators but are designed within the limits of current materials and components. In addition to the gross aspects of anatomy there is the robotic counterpart of neurology: sensing, perception, and controls. Without sensor processing capabilities we have only an automaton. Lacking these, no matter how physically perfect the outward appearance, it cannot be a complete thinking machine that interacts with its environment. These functions may be tested in three ways: (1) Communication imitation test; can a human tell if he is communicating with a computer or another human? (2) Physical imitation test; if you shake hands with an anthrobot does it feel like a human hand? Creating close approximations that can fool humans is not the same as creating an actual living creature, but such tests do measure functionality. (3) The ability of the anthrobot to function autonomously, independent of human intervention, not just reacting to human inputs, places it at the highest level of human behavior. This demands processing of myriad environmental and body image data. Is that my inventor walking towards me? Should I rise and shake his hand? What is he saying and how shall I reply? These thoughts can be generalized as follows.

---

**The Three Measures of Anthrobotics**

Dexterity equivalent to the human

Communication on a human level

Capable of autonomous behavior

---

In short, an anthrobot receives and transforms information into judgements and actions. Its processes should be transparent to human beings, and its outputs—speech, movement, dataprocessing—should be equally transparent.

Dexterity on a human level requires design and construction of anthrobot anatomy with highly integrated kinematics, structure, materials, actuators, and power consumption. The acid test of system integration is how the autonomous device responds to change at a local level, because changes in an integrated system cascade throughout the system. For example, a change in the wrist may result in a need for a less dextrous shoulder or a reduced number of fingers.

Communication at a human level and autonomous behavior require integration of sensors and artificial intelligence to be used effectively. A power source, actuators, joints, and one or more coordinating controllers are required. Artificial Intelligence, perhaps the greatest technological challenge of all time, is advancing with concepts such as neural networks and "fuzzy logic." The processing capabilities of today's robotics beg for greater mechanical dexterity to utilize their processing time fully. Blind robots, through addition of machine vision and other sensing techniques, may actually become lighter with less rigid structures. The high precision that rigid structures provide is unnecessary when you are able to sense and make appropriate adjustments. New structural concepts may be derived, similar to human bone, which breaks and heals (exhibits fault tolerance) rather than overdesigning with consequent increases in mass and weight.

Walking robots will need lighter materials and must be able to recover from falling. Humans and other animals, as highly developed as they are, still break bones. What Frank Lloyd Wright said about architecture also applies to anthrobot design: "form and function are one" (Gill 1987). The blending of form and function is required for a human-like, balanced and integrated design. The failure of so many multimillion dollar robotic efforts can be traced to the absence of high levels of integration. This is symptomatic not only of poor background knowledge of robotics, but also of a lack of understanding about human design. Modularity is often over-emphasized by those who do not understand the value of integrated systems. Well integrated systems must be modular, but only to an extent. Making a single component capable of different functions may create more problems than it solves. For example, each joint in the human body could be of the same design, but the body's joints have different functions and thus by their nature cannot be made identical or modular in the greatest extent.

In the following pages I discuss anthrobotic design considerations, followed by a survey of existing and proposed man-like robots which are the forerunners of anthrobots.

## Musculoskeletal System

The success and value of robots modeled on the human form is self-evident. Man has proven himself as the most versatile and successful animal in the environment. Many researchers have advocated robots based on other life forms such as snakes, spiders, squids, etc. What they do not realize is that these concepts are already subsystems in the human body. For example, the snake's design is found in the human spine, while spiders have a multiplicity of simplified legs to compensate for the low individual dexterity of each leg.

The human musculoskeletal system contains more than 600 muscles and 206 bones in a mature body (Figure 6.1). Actuation of the skeleton comes from large bundles of muscle fibers connected via a seamless transition of tendons and ligaments. Muscle fiber is a universal design

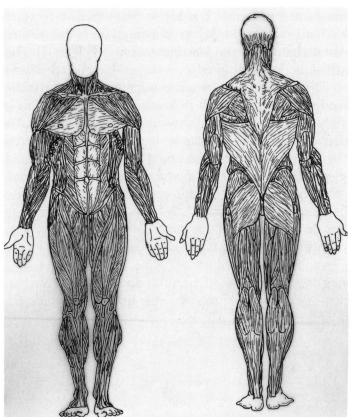
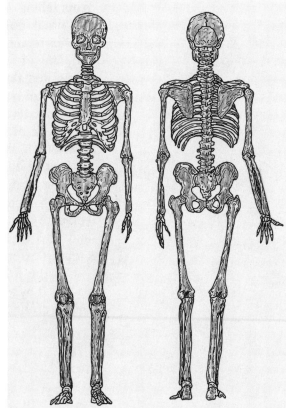

*Figure 6.1  Human Musculoskeletal System*

found in all mammals from shrews to whales. The modular organic nature of muscle fibers lends itself to highly differentiated and customized designs by simply adding more or less fibers or changing the fiber lengths. Different muscle types have evolved for both speed and endurance. Muscles are actuated by electrical potentials generated via nerves that contract a muscle over its entire length at a constant velocity (Hollerbach and Hunter 1991).

A framework of articulated, jointed bones supports the body. Some interesting general observations may be made about the limbs. The arms and legs have a similar basic structure: a three degrees-of-freedom hip corresponds to the shoulder, the ankle corresponds to the wrist, and the feet to the hands (Figure 6.2). However, because of lessened need for dexterity and a higher need for strength and ruggedness, the leg joints are of a simpler construction.

These two subsystems are jointed together by the highly flexible spine. Actuated by muscles located along the trunk of the body. The ribcage connects to the spine in a flexible design that performs the dual function of supporting the viscera and guiding the shoulder blades.

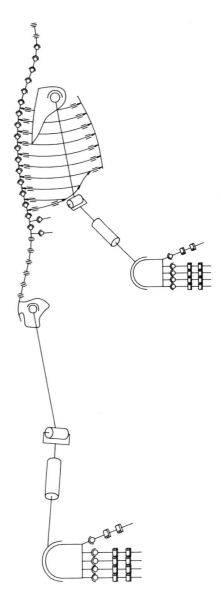

*Figure 6.2 Skeleton Kinematics*

## Materials—General Philosophy

Design of the human body is heavily influenced by the materials composing it. Designers should emulate at least in spirit, this harmonization of bone and sinew that has produced such a beautifully balanced and integrated system. This is not to say that a literal translation is desired; as seen in previous chapters this has not always been successful. Rather, the concept of balance with the materials and components that are available is sought. Features not possible in human systems, such as continuous rotary joints, may be created. By virtue of application some robots must have nonhuman attributes which allow them to function in highly specialized, repetitive, or hazardous environments not suitable for humans.

Anthrobots have special needs reminiscent of aircraft designs. By this I mean that anthrobots, like aircraft, must strike a balance among performance, maintenance, and cost concerns. The two components of cost are acquisition cost and maintenance cost. Acquisition cost is affected by weight and performance factors. Weight must be kept to a minimum while maintaining maximum strength. Computer simulations are not intended as a replacement for actual laboratory and field testing, but should help to minimize these as well as any unpleasant surprises that may occur in testing. Whether an arm, wing, or an entire structure, optimization demands that the designer minimize the negative effects of inertia imposed at the cost of increased weight. The current generations of robots are primarily blind and require structural members of maximum rigidity to achieve accurate repeatability. Anthrobots, on the other hand, cannot have the luxury of rigidity because of the need for reduced weight. The designer will have to turn to other means of optimizing performance such as interactive vision systems and feed forward techniques that can anticipate how much a limb may deflect under load. In the case of moving limbs the further the structure is from its first center of rotation, the lighter and less massive it should be.

The performance issue framed by this mass/inertia issue can be addressed via the use of space age materials. Limited as we are to working with nonliving materials, composites such as graphite epoxy, magnesium, titanium, ceramics, and alloys offer potential advances until self-restoring, fracture tolerant materials are developed in the future. A study of materials was conducted by Rivin in his book *Mechanical Design of Robots* (Table 6.1). Although yield strength is high in many materials, increasing elastic or Young modulus cannot be obtained by heat treating methods of standard materials. Rivin points out that only sintered tungsten alloy has a substantially lower stiffness-to-weight ratio than steel. Beryllium, although mentioned as high stiffness, lightweight metal, is prohibitively expensive. Ceramics, with their potentially low material cost, seem the best hope, if they can be made less brittle and easier to connect with non-ceramic parts. Japan has led the way by initial application of ceramics to small, precision robots (Rivin 1988). Alloying aluminum with lithium is preferred to standard aluminum, but is two to three times as expensive. A carbon fiber reinforced epoxy tube has been used in the ABB GRI OM-5000 forearm, but has not been viewed as a significant factor in improved performance.

The tensile strength of steel is much higher volumetrically than that of tendons; however, its tensile strength per unit weight is only two-thirds that of tendons. In another example, the ratio of Young's modulus to density (a measure of stiffness) for steel is approximately three-fifths that of aluminium. Stiffness is further enhanced by the human body's brilliant engineering that incorporates complex shapes and materials (Todd 1985).

*Table 6.1 Robotic Material Yield and Ultimate Strengths*

| Material | $E$, $10^5$ MPa | $\gamma$, $10^3$ kg/m$^3$ | $E/\gamma$, $10^7$ m$^2$/s$^2$ |
|---|---|---|---|
| Boron carbide, BC | 4.50 | 2.4 | 19.0 |
| Beryllium, Be | 2.90 | 1.9 | 15.3 |
| Alumina, AlO$_2$ | 3.0 - 4.0 | 3.7 - 3.8 | 7.9 - 11. |
| Lockalloy (62% Be + 38% Al) | 1.90 | 2.1 | 9.1 |
| Titanium carbide, TiC | 4.0 - 4.5 | 5.7 - 6.0 | 7.0 - 9.1 |
| Tungsten carbide, WC | 5.50 | 16.0 | 3.4 |
| Molybdenum, Mo | 3.20 | 10.2 | 3.0 |
| Steel, Fe | 2.10 | 7.8 | 2.7 |
| Titanium, Ti | 1.16 | 4.4 | 2.6 |
| Aluminum, Al | 0.72 | 2.8 | 2.6 |
| Magnesium, Mg | 0.45 | 1.9 | 2.4 |
| Tungsten, W + 2 to 4%, Ni, Cu | 3.50 | 18.0 | 1.9 |

Another option is highly engineered materials. Borrowing from the F-16 fighter program, an aerospace construction technique is used in the Thermwood spray finishing robot. "The two lower arms consist of end spars machined from extruded aluminum, with expanded aluminum honey comb bonded between them. Aluminum stress skins are bonded to the outside of the end spars and also between the end spars across the honeycomb (Stauffer 1985)."

Another Thermwood innovation is application of low-cost oil-impregnated bushings set into the structural elements through a special bonding system. Bushing alignment with the shafts is made before the final bonding occurs.

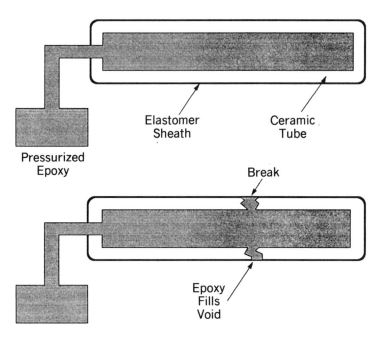

Elastomer Sheath

Ceramic Tube

Pressurized Epoxy

Break

Epoxy Fills Void

*Figure 6.3  Self-Healing Limb Concept*

My concept of a self-restoring or "healing" structural member is shown in Figure 6.3. A hollow ceramic or composite tube housed in an elastic sheath is fed a cement type mixture under pressure. In the case of fracture, the void and the exterior area of the tube is filled with cement, which is cured by chemical interaction with the elastic tube or with catalytic material contained within the elastic tube and exterior surface of the rigid tube. The elastic tube also may be made transparent to allow ultraviolet light curing. Light emitting splints may be used to support the member during the healing process. Another option is a structural member made from shape memory alloys that could be activated by applying heat. Still another concept is pressurized steel tubes for limbs.

As can be seen from this brief discussion of materials, cost can be a limiting factor that forces the designer to make do with aluminum, steel, and plastics, NASA has experimented with pressurized elastic tubes for space station truss members. In cases where a part does not undergo significant deflection, as in the case of the housing and cap of the Omni-Wrist, aluminum or magnesium may be used effectively. Effectiveness of these materials can be enhanced by application of finite element analysis (Feyerabend 1988, 1989). Reduction of weight and optimization for strength are the primary goals. These features produce a faster and more precise design while at the same time reducing material costs. Reasonably priced computer software is now available for performing finite element analysis to determine weight, deflection, and tensile strength. A practical example of this from my own experience is the outer gears in the Omni-Wrist, in which optimization of the design sought to minimize deflection. Adding material to the tracks was validated. Figure 6.4 displays stress measured in psi and Figure 6.5 displays displacement of the track loaded by the roller.

Third, the environmental impact on materials of corrosion, outgassing, thermal conductivity, and temperature endurance become critical in hostile environments such as space where large temperature deviations occur and vacuum conditions exist. Design considerations include making use of rolling contact devices such as ball bearings and use of the simplest form of gearing that will hold lubricants. Lubrication is a critical design consideration, and special wet and dry lubricants such as Braycote or Tribolube may be considered. All imply logistics (repair and maintenance) issues that will have to be addressed on an application by application basis.

At the opposite extreme, undersea applications encounter high pressures and corrosive salt water, which can also short circuit electrical devices. This requires special considerations in materials beyond that of standard stainless steels and plastics.

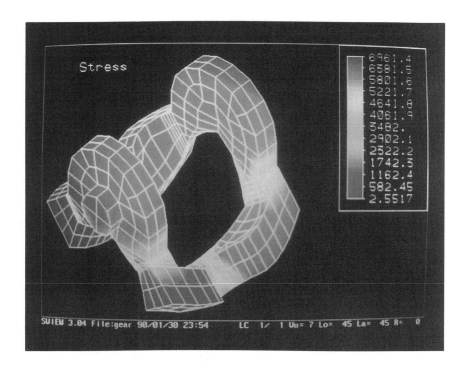

*Figure 6.4  Stress Analysis of Component*

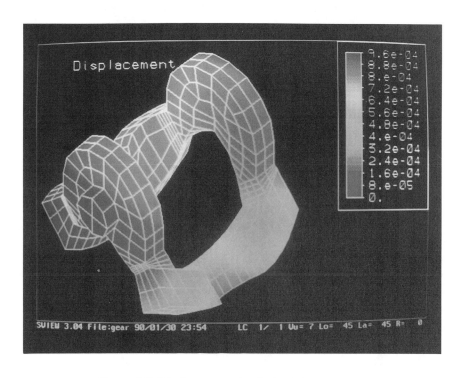

*Figure 6.5  Displacement Analysis of Component*

## Power Systems

The human body's self-contained and protected power supply combined with its fuel acquisition capability makes for a highly mobile and versatile system. The power supply system for the body is provided by the internal organs: heart, lungs, kidneys, etc. Contained within the chest and abdominal cavities the viscera consist of several subsystems, including the respiratory, digestive and urinary systems, used for fueling the body and removing waste. These subsystems find their anthrobotic analogues in power regulators, charging circuits, electrical storage devices and lubrication systems.

Corresponding to the function of the viscera in the human, power systems provide the robot with energy for motion, computing and sensing. Robots "eat" electrical or mechanical energy and convert it into motion or other functions. The efficiency and rate of power consumption is analogous to the metabolic rate in humans. Perhaps a robot dynamometer will someday be invented to "check-up" on the health of robotic patients. Power input, storage, and retrieval are the three functions necessary for the autonomous robot. Of the three, storage remains the area needing the most improvement. Battery technology, the most practical means of power storage, still falls short of the power density needs of locomoted anthrobots.

Lightweight power sources will become increasingly important as robots continue to move away from stationary positions and umbilical connection via power cords and evolve towards self-contained power that allows autonomy and mobility. A growing market for mobile robots may be provided as a spinoff of the development of electric automobiles. The initial market for mobile robots is too small to justify the large research and development required to develop better power sources. However, internal combustion engines have been used successfully in walking robots but are impractical for the larger markets where noxious fumes and noise would be prohibitive. Flywheels and hydraulic accumulators have limited storage capacity but have been used in conjunction with these internal combustion engines as transmission elements. Electrical power remains the most promising. Unlike petroleum powered hydraulic systems that must also carry generators and batteries, electrical power is clean, efficient, and is the same energy form required for the operation of motors and computers. Two of the most promising electrical power systems are batteries and fuel cells. Solar cells might play a role in the far future for trickle charging of batteries. The well turned out robot of the future may carry an umbrella covered with solar cells, popping it open for a quick pick-me-up.

The first rechargeable lead-acid battery was invented by Raymond Gaston Plante in 1859. Other types, nickel cadmium for example, have superior cycle life and deep discharge capabilities, but are several times as expensive. Lead-acid batteries are the most popular type, accounting for over 60 percent of all battery sales. The three major lead-acid batteries are flooded, sealed nonrecombinant (used by most automobiles for starting), and sealed recombinant. They all share the same electrochemistry found in a lead dioxide/sponge lead cell (Meurer 1990). Sealed recombinant lead-acid batteries (Figure 6.6) are used by TRC for the Helpmate robot. This type of battery was chosen for its lack of outgassing and the spillproof construction necessary in the hospital environment (Weber 1991).

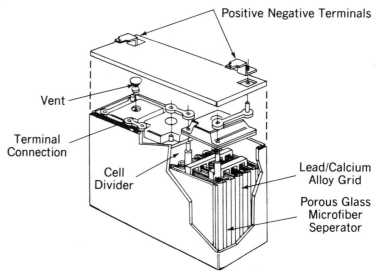

*Figure 6.6  Sealed Recombinant Lead-Acid Batteries*

 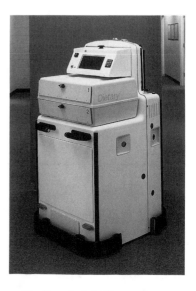

*Figure 6.7  TRC Helpmate Powered by Lead-Acid Batteries*

Lead-acid batteries provide the Helpmate (Figure 6.7) with up to ten hours of operation with a life of 300 cycles. A concern is to neither allow the battery charge to drop below 80 percent of capacity, nor to recharge within 24 hours. Both events damage the battery, much like underfeeding or overfeeding damages a human. Also important is to charge the battery fully before its next use (Weber, personal communication 1991).

## Power Distribution Systems

Over 60,000 miles of tubing comprise the average human circulatory system. Blood sends fuel to the limbs and organs and removes waste products from these areas. The heart provides the system pressure that sends freshly oxygenated blood via the aorta, the body's largest artery. The system branches out to arteries that in turn branch into millions of microscopic capillaries which run into every corner of the body. The wire harness for the anthrobot's electrical power system would be analogous to the human circulatory systems. Electrical power flows via the wires to motors in a circuit, much as arteries and veins are routed in the human. The one-way valves in the veins are analogous to surge suppressors and other electrical regulatory devices needed to protect an electrical system. The very high flexibility of arteries and veins poses a challenge to the robot wire harness designer. Tight radii and high angulation joints tend to kink and abrade wires. Super flexible wire, such as that used in the Omni-Hand, which is composed of finely woven shielded wires would provide a reasonable equivalent to arteries. Standoffs or other provisions for wiring also must be integrated. Passages inside a joint may reduce the bend radius and at the same time protect the wiring. Protecting the wire through skin-like materials is another alternative. Materials developed for space suits, such as high flexibility, Kevlar/Nomex Teflon coated blends, that incorporate a special weave to prevent tearing, are another option.

## Fuel Cells

First described in 1839 by Sir William Grove, fuel cells had to wait for NASA's need for a high power-to-weight ratio before finding a practical application. A form of continuously operating battery, they provide a lightweight, high power density alternative to batteries. The fuel cell is a highly efficient power generating system that produces electrical power by the electrochemical conversion of hydrogen and oxygen into water and electricity (Figure 6.8). The anthrobot would thus eat hydrogen and oxygen gas, converting it into energy. Consisting of a multilayered sandwich of anodes and cathodes separated by an electrolyte, fuel in the form of hydrogen and oxygen is pumped in. Hydrogen diffuses through the anode and reacts with the electrolyte to produce electricity. Water and heat are produced as by-products, much as in the human system. The circuit is completed when electrons return to the cathode, where they are picked up by the supplied oxygen (Fisher 1991).

### How a Fuel Cell Works

Fuel cells convert chemical energy to electricity. Because there is no combustion, these cells give off few emissions.

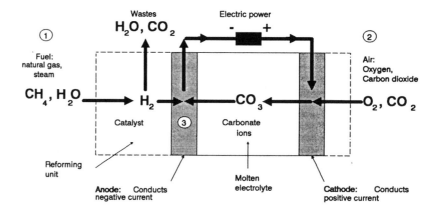

① Fuel passes through the reforming unit, which extracts hydrogen gas.

② Air is pumped into the chamber. Oxygen and carbon dioxide pass through the cathode and react with the molten electrolyte to form carbonate ions.

③ When hydrogen reacts with the carbonate ions it splits, giving off a stream of electrons or electricity.

*Figure 6.8  How a Fuel Cell Works*

There are five types of fuel cells: Alkaloid, Phosphoric Acid (becoming popular in power plant applications), Solid Polymer (undersea applications), Molten Carbonate (future powerplants), and Solid Oxide (future powerplants). Energy density of 150 Lbs per KW/hr for a conventional lead-acid battery compares favorably to 5 Lbs per KW/hr in a fuel cell. Fuel cell cost is still very prohibitive; 1 KW costs $576,000. Alkaloid fuel cells used on the NASA space shuttle are one-tenth the weight of the best available battery technology and are well suited for use in anthrobots. Perhaps future anthrobots will benefit from spinoff advancement brought about by space, powerplant and electric automobile applications (Woodruff 1990).

We now turn to other key anthrobot components - sensing/feedback organs. They provide an interface to the anthrobot's environment. Without them the anthrobot is deaf, dumb, and blind.

# Sensing/Feedback Organs

The "five senses" of the human are sight, hearing, touch, taste, and smell. A sixth sense—the vestibular system located in the inner ear—also exists and provides us with a sense of balance, knowledge of position in space, and when combined with other senses, body image. The human senses gather information that allows us to interface with the environment. Touch on a finer level provides position sensing of the limbs and input about environmental conditions (temperature, humidity, etc).

Functions of the anthrobot sensing system organs are information gathering about the environment and interfacing with the environment. A subset of this is position sensing of the robot itself and its associated limbs. Integration of sensing systems requires a realtime sense of position, attitude, movement, location of parts with respect to the whole body and location of the body with respect to environment.

The interface of anthrobot and environment is the most critical place where errors might occur. For this reason sensing must be not only accurate and rugged, but also recoverable. The interface is crucial to the anthrobot's ability to affect its environment, so the quality of the interface with respect to data collection or speed of transfer is crucial to performance. On the other hand, optimization of anthrobot functions by specialization of its function would reduce the need for full humanoid sensory integration. The anthrobot would only need enough capability to perform in a dedicated function.

## Hearing

The function of hearing is to provide a communication interface that provides the location of objects, or in the military parlance—target acquisition. Mounted in pairs on the sides of the head, ears are capable of three degrees-of-freedom. Interestingly, the shape of the ears provides added direction sensing. This arrangement provides auditory data, from speech to other audio signals, which allows direction finding. The human ear has an auditory range of 15 to 15,000 Hz. A complex linkage mechanism transmits information to nerves that forward the information to the brain. Obviously, microphones are the closest equivalent to the human ear and are much simpler. Their frequency range is also broader than that of the human ear, typically 10 to 100,000 Hz.

**Sonar**

One of the applications of human hearing, besides speech recognition, is to guide a person toward or away from an audio signal. However, human hearing requires an object to emit a sound. Sonar, on the other hand, has been developed by bats into a self-contained system that transmits a high frequency signal from the mouth. The signal reflects off objects and is received by their ears. Thus the target or object does not have to emit a sound for it to be tracked with sonar.

Triangulation is a basic navigational technique used in sonar, radar, and optical systems to determine position of a object. A beam is transmitted from two sources and the angle of the beams reflected off the target are measured. The target's position relative to the sensor can thus be calculated.

The TRC Helpmate robot (Figure 6.9) uses an array of Polaroid ultrasonic rangers to detect and measure the distance of obstacles or other objects during navigation. Range is from less than 1" to 35". Accuracy resolution is plus or minus 1/8" to 10', ±1% of the reading to 35". The ultrasonic rangers were originally developed for Polaroid instant cameras but have now found multiple uses, including anthrobotics.

The Polaroid Ultrasonic Ranging System uses a bidirectional electrostatic transducer. It generates a short burst of high frequency sound, which travels out from the transducer and is reflected back to the transducer by objects in its path. The transducer receives this reflected sound energy and converts it into an electrical signal. The distance from transducer to target is then computed by additional circuitry.

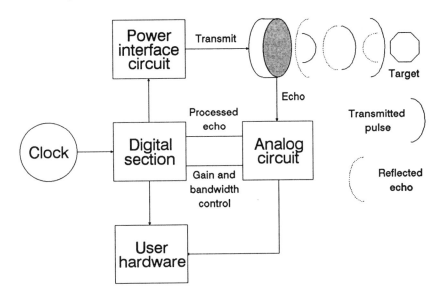

*Figure 6.9 Ultrasonic Ranging System Block Diagram*

A variation of basic sonar, doppler sonar distinguishes a moving target by the phase shift of the sound wave as it is returned from the target. This is useful to determine speed as well as for differentiating between multiple objects which may or may not be in motion.

## Radar

Radio Detection And Ranging (radar) was first suggested by Nikola Tesla in 1900. He postulated that through application of electromagnetic waves, the relative position, speed and course of a moving object could be determined. In 1934 at the Naval Research Laboratory Washington, D.C., Leo Young and Robert Page received the first radar echoes — saturation signals from a "wood and fabric" airplane at one mile. Research and development continued with installations beginning in the 1930's on battleships.

Like sonar, radar transmits a signal that bounces off an object. The echo is received and processed into position information. Microwave signals in approximately the $10^{10}$ range are broadcast from a rotating transmitter. Since the transmitter is "off" between pulses, the same antenna can be used to receive the pulse echoes. Range and direction of the object that originated the reflection may be determined. The choice between sonar and radar depends on environment and application; radar is good for locating large objects at a distance while sonar has more granularity but shorter range, and can operate underwater.

Infrared sensors may be made to function like radar, but only in the infrared frequency range. A light emitting diode transmits a narrow beam and is received by a phototransistor. Range is up to 30 feet, but knowledge of the surface refractivity is needed to accurately determine range, better suiting infrared sensors to determining target angle. Triangulation techniques may be used to determine distance if the beam is transmitted at an angle, a method commonly used in cameras. Infrared may also be used for passive detection more like human hearing. An infrared camera can "see" a human because of its higher body temperature versus background temperature.

## Inductive Sensors

These sensors have the nonhuman ability to detect ferrous or nonferrous metallic objects (Figure 6.10). They operate on the principle that magnetic or conductive surfaces entering an electric field cause distortions detectable by the sensor. This leads to the interesting range of senses outside of human experience. An anthrobot so equipped might "feel" an automobile by its magnetic field without actually hearing or seeing it. Consisting basically of a coil and ferrite core arrangement, an oscillator and detector circuit and solid-state sensor. The oscillator creates a high-frequency field, radiating from the coil in front of the sensor, centered around the coil's axis. The ferrite core focuses and projects the electromagnetic field forward. When a metal object enters the high-frequency field, eddy currents are induced in the surface of the object. The net result is an energy loss energy in the oscillator circuit and thus producing a smaller amplitude of oscillation. Range is from approximately 2 to 20 mm in sizes up to 30 mm in diameter. Larger "pancake" sensors provide greater sensing range up to 20 cm. The analog signal is proportional to the distance of the target and may be used in servo loops or to trigger an output at a set distance.

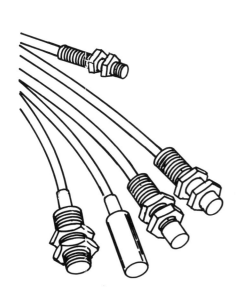

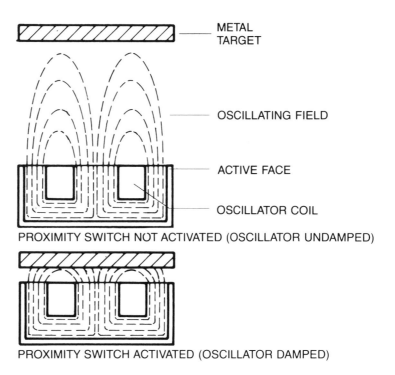

METAL TARGET

OSCILLATING FIELD

ACTIVE FACE

OSCILLATOR COIL

PROXIMITY SWITCH NOT ACTIVATED (OSCILLATOR UNDAMPED)

PROXIMITY SWITCH ACTIVATED (OSCILLATOR DAMPED)

*Figure 6.10 Inductive Sensors*

**Taste and Smell**

The sense of taste provides an analysis of solids or liquids into four primary categories: sweet, sour, bitter, salty. In the human, ingested molecules dissolve in saliva and bud shaped clusters of cells, taste buds, located in the lining of the tongue's papillae analyze the chemical signals of the food-saliva mixture and transmit the information through nerves to the brain. Taste provides a nonvisual and nonauditory means of substance identification.

The anthrobotic analogue of taste may include chemical analysis of various materials and liquids on earth or planetary soil analysis. The anthrobot may take a sip from a stream or "eat" some soil and provide a breakdown of its contents. This has been done on the Viking landers. The soil samples were "tasted" by miniature automated laboratories. The laboratories included a gas chromatograph, mass spectrometer processor and a biology processor which added chemicals and monitored the subsequent chemical reactions.

The human sense of smell breaks down odors into four primary groups: fragrant, acid, rancid, and burnt. Figure 6.11 shows a sensor design for detecting carbon monoxide in flue gas emissions. Commercial sensors are available that select only one type of gas such as carbon monoxide and filters out other gasses. These electrochemical sensors have an extra electrode that measure other gasses so their value does not interfere with the primary gas measurement. A robot so equipped could be used to sniff the emissions in smokestacks or the atmosphere on Venus.

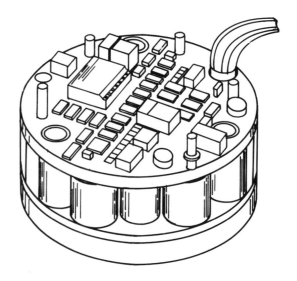

*Figure 6.11 Sensor for Detecting Carbon Monoxide Gas*

**Touch**

Functions of the sense of touch include collision avoidance and modulation of pressures exerted by endeffectors, as well as position and temperature sensing. Machine sensors may include not only tactile, but pressure and temperature sensors to detect undersea, atmospheric or other types of pressure, including many outside the human range of sensing.

The surface of the human body is covered with 6 Lbs of skin having an average thickness of .20 inches. Within this narrow area is found a large network of nerve endings. The human hand, for example, has up to 1,300 nerve endings per square inch. The fingertips are twice as sensitive as other areas of the hand and more sensitive than any other part of the body, excluding the lips, tongue and tip of nose. This sensitivity allows the fine feedback and control needed to perform the manipulations characteristic of the human hand. Measuring pressure, temperature, and pain caused by abrasion or puncture, it is also able to identify varying degrees of moisture, which is a mixture of coldness and skin surface sensation.

Force-torque sensors are used commercially to approximate the sense of touch. Quoting from JR3, Inc. literature: "By measuring the forces on three orthogonal axes, and the moment or torque about each of these axes, a six degree-of-freedom sensor can completely define the loading at the sensors location (Figure 6.12.). Use of six degree-of-freedom sensors in robotic systems allows adaptive process orientation."

*Figure 6.12 Six-axis Force Sensors*

Machine tactile sensing was once considered to be a large market, but robot tactile sensor companies have all but disappeared—out of touch with the market realities. Companies still making tactile sensors have survived because of other market niches. This situation is probably the outcome of a robot market still dominated by demand for relatively gross motion and limited need for fine positioning. In the future, anthrobots, freed of repetitive tasks and roaming at large in a complex environment, will require more sophisticated sensors and consequently tactile sensing will experience a resurgence in demand.

Tactile sensing is also used for short range proximity sensing. Microswitches are a crude version of a single nerve. Bumpers incorporate microswitches for fail-safe collision protection in Autonomous Ground Vehicles (AGVs). Constructing them into arrays or attaching them to feelers, like a cat's whiskers, further enhances and extends their information gathering power.

Tek-Scan sensors were originally created to fill the need for an electronic dental impression system. Figure 6.13 displays a sensing cell detail. This has been customized for the Omni-Hand described in Chapter 4. Highly flexible and inexpensive, it is made by screen printing the conductive grid on two sheets of polyester film, to which is added the resistive layer, insulation where required, and glue to hold the sheets together. The Tek-Scan sensor is a grid based device. The sensing surface consists of conductive rows and columns whose intersection points form sensing locations. Separating the rows and columns is a material that loses its electrical resistance with force, and by doing so each intersection becomes a force sensor.

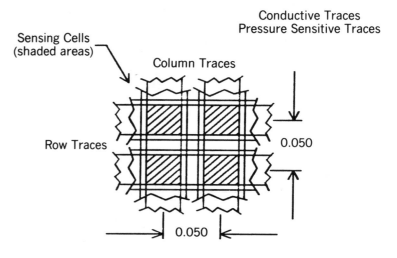

*Figure 6.13 Tactile Sensor by Tek Scan*

Another type of tactile sensor is made by Interlink Electronics (Figure 6.14), which developed sensors originally for electric pianos. The Interlink Electronic system operates on the principal of a force sensing resistor. Featuring a wide impedance range, it is a "polymer thick film device which exhibits a decreasing resistance with increasing force applied normal (i.e., down) to the surface (Interlink)." Several sensors are linked in parallel with a single output. These sensors have been used successfully in the Belgrade/USC Hand described in Chapter 4.

Tactile sensing arrays are being developed by the University of Utah that reduce cabling by utilizing multiplexing techniques. Hybrid preamplification and multiplexing circuits mounted on the back of the fingers have led to a sophisticated sensing system of 677 capacitive transducers. A large area is covered and arrays are located in the palm, fingertips, and links (Jacobsen 1987, Venkataraman & Iberall 1990).

Tactile sensors are important for supplementing information from the vision system. Knowledge about surface texture and composition is valuable when processing received signals. For example, a full-scale image in a mirror should be interpreted differently than the actual item. Surface reflectivity also can be used to calibrate vision systems that use shape from shading algorithms. An anthrobot actually may stroke an object to determine if its surface is shiny and metallic as a double-check for misinterpreted visual cues (Engelberger 1989).

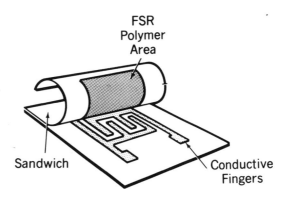

*Figure 6.14  Tactile Sensor by Interlink*

## Body Image

Functions of body image data include balance and position sensing as well as a sense of self relative to space. The joint angle sensor provides proprioceptive data and is one means by which the anthrobot may build an image of itself. Encoders, resolvers, and potentiometers have been traditionally used for determining joint position. Figure 6.15 displays an optical encoder. A glass disc is covered with opaque sectors representing 0 or l. Each segment of the disc is encoded with a unique binary signature that provides absolute positioning which is recoverable even after a loss of power. Light passing through the disc is read by optical sensors that provide the computer with position information through the servo amplifier. The greater the number of lines on the encoder, the greater the resolution. Other types of position feedback devices include magnetic, in which a magnetic disc is encoded with lines of force. Like the optical encoder, these are read by a sensor, in this case a magnetic pickup. Frameless resolver feedback is another type where the sensor consists of a coil of wire rotating inside another coil.

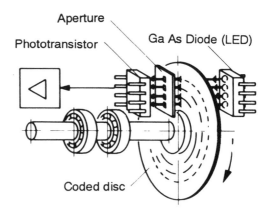

*Figure 6.15  Joint Angle Sensor*

In the inner ear, the dedicated mechanism of the semicircular canals provides position information to the human brain for balance. They help the human body maintain an upright position and supply information about the body's angle relative to a ground plane.

Some of the limitations of the human system have been discovered by the aviation industry. Pilots without visual cues to true vertical become disoriented. Truly pedestrian, humans are also poor at sensing a rate of turn or remembering true vertical.

The gyroscope, commercially developed at the turn of the century for compasses and aeronautical instruments, is the mechanical equivalent of the inner ear and was among the first servo mechanisms. Figure 6.16 displays a schematic of a typical two degrees-of-freedom gyro in its gimbals. It provides a measure of angular displacement of the case about the two orthogonal axes. Tilting of one gimbal axis produces torque, known as precession, at the perpendicular axis. Pickoffs at the gimbal pivot points provide a reference signal to indicate the gyro's angle in space.

A well designed anthrobot navigation system will require more than a gyro. Gyros suffer from a problem known as "drift." Although the gyro provides very accurate attitude information for a short period of time, drift is introduced by many incidental factors that ultimately produce inaccuracy. Therefore, gyros are used in conjunction with vertical indicators that are not affected by drift, such as electrolytic sensors that function like a carpenter's level. This information can also be used to create an inertial guidance system by integrating a clock. Distance as well as direction can then be calculated. Systems like this are widely used in missiles and aircraft. Besides the classic spinning mass mechanical gyro described above, other more recent types include reed, gas rate and laser gyros.

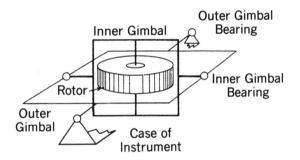

*Figure 6.16 Schematic of Two Degrees-of-Freedom Gyro*

The "sixth sense" body image gives the anthrobot the feel of personal orientation consisting of its overall shape and location in space. This is valuable to detect intrusions by objects or people and to provide knowledge of location and orientation. Capacitive proximity detectors may be used to generate a detection field. A radio frequency antenna radiates an electromagnetic field, and if a human or other dielectric intrudes, a capacitance to ground is produced. This is detected by the system's radio frequency oscillator. Nondirectional, the range is two feet or less for objects with the mass of people. Field shape is a function of antenna design and ground plane shielding. Sensitivity thresholds should be set at a level that doesn't detect the movement of arms or other limbs. On the subsystem proprioceptive level, joint angle sensors such as tachometers, encoders, and potentiometers can provide localized body image data. In "Micropositioning Apparatus for Robotic Arm" (Bartholet 1989) lasers are used to detect deflection and provide for precision micropositioning of robotic limbs (Figure 6.17). An array of light emitting sources at one end of the limb transmits light to an x and y axis sensor. Any deflection of the limb is measured and that information may be used for position correction. This allows for lightweight and therefore less rigid structures while still maintaining precision. Satellite navigation systems could be integrated for a broad sense of location in the case of anthrobots that traverse long distances.

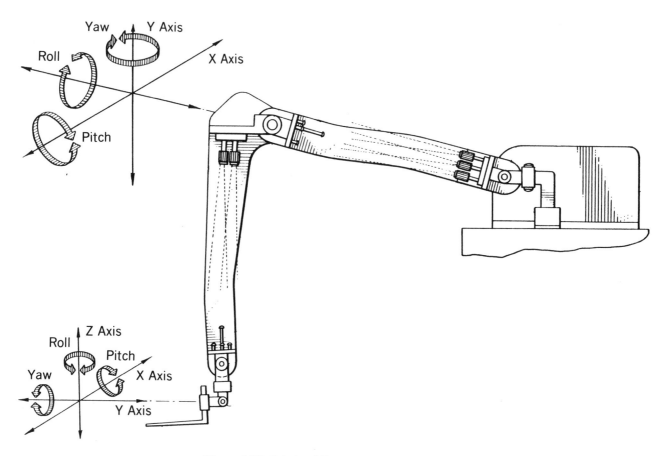

*Figure 6.17 Odetics Micropositioning Apparatus*

## Odometry

Odometry is an ancient method of recording distance. Recorded by Hero and Leonardo da Vinci (Figure 6.18), odometry offers a low cost, reasonably accurate method of providing proprioceptive data by measuring the number of wheel revolutions and extrapolating distance from that data. However, gradual drift occurs from wheel slippage. Used in conjunction with a home position, beacon or other aid, this method offers cost effectiveness. Odometry has been used in conjunction with an electronic compass by KVH Industries Inc. for electronic automobile navigation systems. Turns are used to constantly recalibrate the system. In the Intelligent Vehicle Highway System (IVHS) system (Figure 6.19) being developed by Motorola and the University of Illinois at Chicago to ease congestion in the Chicago area, both dead reckoning and Global Positioning Satellite System are used to determine vehicle coordinates. In this IVHS system, "Advanced Vehicle Advisory Navigation Concept" (ADVANCE), is used to provide drivers with dynamic routing information to help the driver choose the best route to his destination. An onboard navigation computer exchanges critical traffic information via radio frequency with central traffic computers. IVHS technology could be spun off to develop anthrobots' tracking and control systems, where the anthrobot would take the place of the vehicle. Entire armies of anthrobots could thus be coordinated and controlled.

*Figure 6.18 Leonardo da Vinci's Odometer*

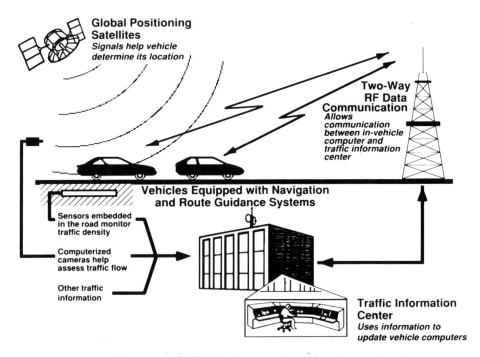

*Figure 6.19  Intelligent Vehicle Highway Systems (IVHS)*

## Compasses

The compass senses the earth's magnetic field and provides a global wide reference. Modern compasses use electronics to provide extremely accurate heading data in either digital or analog formats (Figure 6.20). A saturable ring core, free-floating in an inert fluid is magnetized by two primary coils wrapped around the housing. Two secondary coils measure the amplitude of the induced pulses which are proportional to the earth's magnetic field. Spurious magnetic fields, such as the robot's metallic components or circuits, could be nulled during installation.

*Figure 6.20  Electronic Compass*

## Human Vision

The human eye is a nearly perfect optical instrument. Seventy percent of the human sensory cortex is dedicated to vision. More directional than the ear, the eye possesses its own muscle driven, fine positioning system. Coarse positioning is provided by the three degrees-of-freedom neck. Cushioned against shock and vibration by fat, the eye focuses by changing the curvature of its lens while the iris regulates the entrance of light. With a sensitivity of 1/100,000,000,000 of a Watt, cells in the retina called rods register black and white only. Color is perceived by the cones. Thus, the human vision system operates on two levels: The older, black and white, motion and contrast sensitive system; and the newer, color sensitive system. While less capable of seeing detail than the color system, the black and white system is much more functional in low light situations.

## Machine Vision

The issue of vision is one of the central senses for anthrobots, as much of human functioning is dependent on sight. However, there are many types of sight which are effective. Machines do not have to gather the same data that we do, but they do need to be able to come to similar conclusions. Bear in mind that what is seen and how it's seen are only a part of sight. Processing and interpretation are what make vision so useful. It's very important that anthrobots be able to process whatever raw visual data they have, be it pixels, infrared, scanned images, etc., in such a way that they get the same result that people do. Thus, anthrobots should look at a chair or a cloud and know that's what they are called by people, regardless of how these might "look" to them. Types of machine vision need to be evaluated relative to their ability to promote anthrobotic autonomy and to facilitate man-machine interaction.

## Photoelectric Eyes

Photoelectric eyes simply sense the presence of light with no detail (Figure 6.21). Photoelectric eyes used in shops to announce customers function at this rudimentary level. The light source is either opposite the detector or housed together with it, and an opposing mirror completes the path. In some versions the frequency of the light approaches the infrared spectrum and allows the eye to look through cardboard boxes and other medium density objects.

*Figure 6.21  Photo-electric Eyes*

Beyond the rudimentary techniques described above, the three main classes of visual perception are as follows. Level 1 image processing is a form of signal conditioning. Image data is processed with computationally simple algorithms to enhance the signal-to-noise ratio. Level 2 consists of recognizing patterns of image features such as edges, areas or holes, and their relationships to each other. Most two-dimensional machine vision systems fall into this category, seeing a silhouette, but they require high contrast without overlap. They are used primarily for inspection and gauging applications. Level 3 is the human level of functionality, characterized by the function and the ability to interpret what is seen, to recognize 100 different faces almost instantly.

### Two-Dimensional Machine Vision

Two-dimensional machine vision systems were pioneered at SRI International and Stanford University. Isolated parts like switch plates may be located and inspected. The plate is silhouetted against high contrast lighting, and a camera sends the image to the preprocessor, which reads it for the microcomputer that contains the vision algorithms. Thresholding converts the image into a unambiguous black and white silhouette. Algorithms trace the plate's outlines and extract geometrical information which is compared to known data. The SRI algorithm provides the machine with a list of features that it can use to identify the object. The machine is trained by repeatedly exposing the object and thereby recording all shape descriptions. Thus if the plate is missing a screw hole it will be rejected (see Figure 6.22). A highly commercialized field, there are many vendors offering systems. Limitations of these systems are (1) high contrast required, (2) no overlapping, and (3) slow performance, especially if several geometrical descriptions are required.

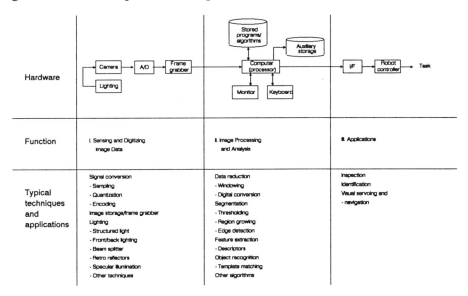

*Figure 6.22 Silhouette Machine Vision Image*

A new approach, the Wilkie, Stonham and Aleksander's Recognition Device (WISARD), is based on neural networks devised at Brunel University by Igor Aleksander (Aleksander and Burnett 1987). Designed to function like the human vision system, WISARD's limitations are less severe than the two-dimensional vision system, and it may even recognize human faces regardless of facial expression. Because the neural network does not require preprogramming it is more adaptable and recognizes a greater range of visual images. WISARD consists of several separate neural networks, or discriminators, that independently may be taught an image. The neural network contains 30K of 8-bit RAM. Each stored portion of the image contains eight bits of information that is "learned" by each RAM. The thousands of rough images recorded by the RAM are summed by the neural networks and displayed as a bar graph. The bar is fed back into the system and mixed into the image being viewed. Recognition is reinforced when this feedback is fed into a net that is learning the image. Image resolution is 512 x 512, or roughly the same as a domestic television. In operation the system functions like the human visual system; it is capable of focusing on details while still maintaining a peripheral view of the surrounding scene. This ability to separate foreground from background levels permits humans to focus on important details while still maintaining overall visual awareness. WISARD provides feedback, an internal self-check much like the human process of recognizing a face, not by a specific outline as in the SRI systems described above, but by a countless mosaic of patterns.

## Foveated Retina Sensor

Developed by the Department of Electrical Engineering at the University of Pennsylvania and the Belgian Interuniversity Microelectronic Center (IMEC), this charge coupled device (CCD) imager (Figure 6.23) is loosely based on the human retina and has advantages over conventional raster scan devices (Van der Spiegel 1989). Embedded into its structure is a logarithmic transformation that makes an image's size and rotation invariant. The CCD imager consists of 30 concentric circles and 64 sensors per circle, whose pixel size increases linearly with eccentricity. The central part has a constant resolution with 102 photocells. The CCD is made in a three phase buried channel technology with triple poly and double metal layers. By emulating the human eye's design a number of machine vision problems are solved. This is in contrast to the previous systems that relied on processing the vision data after it leaves the camera. Like the human eye, photoreceptor resolution is greatest in the center of the visual field and decreases in density toward the periphery. Once the peripheral photoreceptors have detected their subject the eye may be swiveled to focus the higher density center of the retina for detail, which eliminates the need for processing a wide field of view. TRC is pioneering the application of this type of device in its log-polar technology. Figure 6.24 illustrates the image as seen by the human eye. Resolution decreases away from the center of view. That is, the optical image is sensed via an array that is patterned like Figure 6.24d, with fine cells in the center, grading to coarse cells in the periphery, which the sensor in Figure 6.23 emulates. Neural networks in the human retina further process the original image to yield representations such as shown in Figures 6.24b and 6.24c, which provide summarized data over larger cells of area. All three of these representations map in Log-polar fashion into the brain. That is, rings of imagery map into columns. The term "Log-polar" means that the coordinate system in the brain is like polar coordinates, except that the radial component is logarithmic. Log-polar coordinates simplify the task of the brain in interpreting imagery by turning rotation and zoom into simple data shifts (Chaikin and Weiman, 1981). No numerical operations are required to recognized objects at any ranges of size and orientation. This improves the speed of visual recognition and simplifies the required computing machinery. The anthrobot, like its human counterpart, needs realtime coordination between sensory perception and motor control as well as target recognition and tracking. This technology seems particularly relevant.

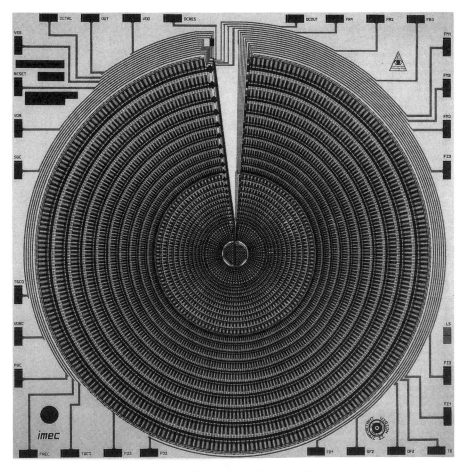

*Figure 6.23  Foveated Retina*

*Figure 6.24  Uniform Resolution Photograph*

*Figure 6.24a  High Resolution Log-Polar Image*

*Figure 6.24b  Medium Resolution Log-Polar Image*

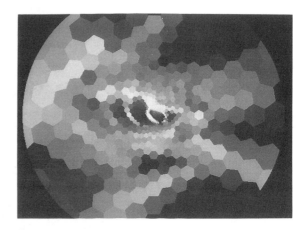

*Figure 6.24c  Low Resolution Log-Polar Image*

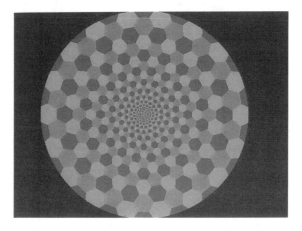

*Figure 6.24d  Hexagonal Tessellation Log-Polar*
*Sensor Array—Analog of Human Retina*

## Three-Dimensional Vision

Binocular stereo vision is a proven method used by humans and other animals to provide the brain with three-dimensional images and range information. However, it requires advances in artificial intelligence before it can be applied to anthrobots. Problems in determining the correspondence of the two images of a single feature result in processing delays and cost penalties. Important for binocular systems are coarse and fine camera aiming systems that are analogous to the neck and eye muscles (Figure 6.25). This simplifies processing of images by mechanically correcting attitudes of cameras to the object being viewed and thus reduces the amount of image processing required to correct for odd viewing angles (Ballard 1989). When we look closely at an object we usually look at it head-on, which reduces the amount of thinking required to process the image.

At a distance, three-dimensional vision becomes range information. Binocular systems use images from two separate viewpoints and triangulate range based on the measured difference in image position between the identical scenes (Engelberger 1989). For robots to function in a variety of unstructured environments, realtime three-dimensional depth is needed. Adding color offers some interesting possibilities that make detection of certain substances easy. A robot surgeon, for example, may be sensitized to look for the color red (i.e., blood) without having to perceive any image, shape or detail.

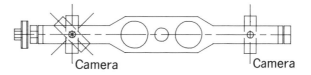

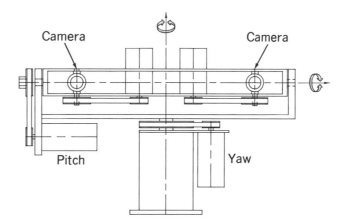

*Figure 6.25  Stereo Vision System*

## Structured Light

One of the earliest techniques to compete with the human three-dimensional vision system, structured light is the most interesting to the anthroboticist. Structured light systems prestructure and limit the amount of information sent to the computer.

In such systems a light grid is projected onto the subject which permits an unambiguous interpretation of edges and surfaces. Shadows and other extraneous effects are ignored by the vision system automatically, greatly simplifying processing. Another benefit is that the projected light geometry is known and thus provides a binary two-dimensional image that contains three-dimensional information. Figure 6.26 displays such a system used for seam tracking by robot welders. The seam is scanned using several strips of laser light. Image information received by a camera is broken down into individual pixels which are processed into depth information by determining their deviation from the original line geometry.

For three-dimensional image processing a more complex light grid is projected over the object and correspondingly more complex video processing techniques are required. When a grid is projected over a sphere, for example (Figure 6.27), the grid appears denser as it extends towards its edges in proportion to the cosine of the angle between the surface normal and illumination direction. This increased grid density provides information the processor uses to determine the three-dimensional shape.

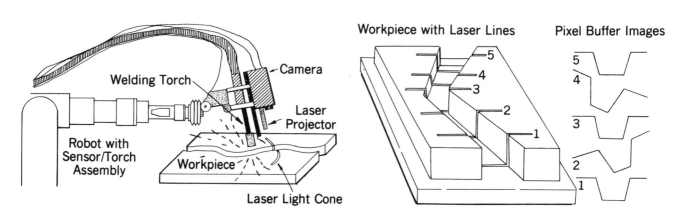

*Figure 6.26 Seam Tracking by Robot Welder*

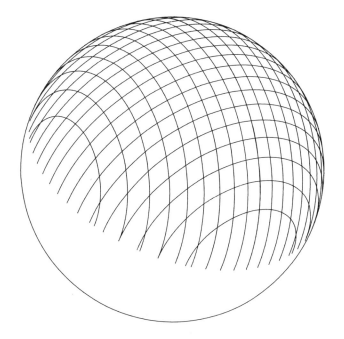

*Figure 6.27 Light Grid Projected on Sphere*

## Laser Time of Flight

The 3D Laser Imaging System by Odetics is an interesting hybrid between conventional machine vision and structured light systems (Figure 6.28). Also known as a 3D mapper, it measures the phase shift of an amplitude modulated infrared laser to obtain three-dimensional data. When the data is seen on a monitor different colors indicate range. In operation, video pixels are scaled to the logarithms of the detected laser power. Resolution is comparable to a black and white TV camera. Since the laser provides the illuminating source the video pixels are independent of ambient illumination. No shadows are produced since the laser light source is coincident with the photodetector. This independence of ambient lighting is valuable in environments where ambient lighting is uncontrolled or nonexistent.

The scan unit measures 9" x 9" x 9-1/2" with a nonreflective glass window and a cover on the front. A box contains the scan mechanism and scan optics, the laser and laser modulator, electronics units and the avalanche photo diode. In operation the modulated laser beam is fanned out in a 60 x 60 field of view by the rotating the drum mirror. As the light bounces off the target and returns to the avalanche photo diode, a phase shift in the modulated signal results from the distance the laser beam traveled. The signal is processed through the electronics unit and the avalanche photo diode to produce range and video images. A two-dimensional image is formed containing three-dimensional data of the target. This device has been used successfully in the NASA JSC Extra Vehicular Activity (EVA) Retriever to provide primary visual information for grabbing tools and accurately sensing a three-dimensional environment.

Improvements necessary to gain wide acceptance of this technology are simplification of the costly rotating mirror drum, weight reduction, and lower power consumption. Prohibitively expensive, the system is also fragile and complex.

*Figure 6.28  3D Laser Imaging System by Odetics*

# Speech

The elements that receive and transmit sound have been the simplest for robotics to duplicate. Microphones and loudspeakers approximate the vocal system and ears of the human in a much simpler manner, requiring far fewer parts than the machine senses. Creation and understanding of human speech is in the field of voice synthesis and voice recognition, and begins to meld into Artificial Intelligence.

## Machine Speech

In the late 1870s Thomas Edison predicted that clocks and vending machines would talk to announce the time and sell products. Not until recent times was this dream made practical with rugged and reliable solid state devices. Synthetic speech is typically created from ROM and RAM codes or from ASCII text. Forty different speech fragments called phonemes have been identified. Their pronunciation is contingent on the possible 128 allophones which come after or before the phonemes. In addition other nuances called prosodic features—inflection, pause, and pitch - are needed to enhance meaning. To construct a word, a phoneme string is selected, pauses are provided, and prosodics are added and further refined through testing. Dictionaries called memory lookups are one computationally intensive method of providing realtime vocabulary. Another more economical approach reconstructs words from fragments called morphemes. Eight thousand morphemes are required for a basic vocabulary. Just as Edison was bedeviled by words beginning with the letter s, crackling sounds, poor recording, and uneven tracking in his development of the phonograph, contemporary speech synthesizers have to overcome the machine-like quality common in today's systems. The future appears bright for the new pioneers who seek, like Edison, to make machines talk.

## Recognition

In the inverse of speech, words must be deconstructed through computer algorithms to be comprehended. There are more variables in understanding spoken words than in creating them. Unlike speech synthesis, which may rely on a stored vocabulary, speech recognition must be real-time to prevent long pauses in conversation. Processing connected speech, not just individual words, is the challenge of today. The words must be understood in their context, not just recognized. The equipment that recognizes words works well enough; the processing capability must now be developed and advanced. Voice recognition is difficult, as the processor must understand a variety of pronunciations by different individuals under unpredictable levels and types of background noise. Will the anthrobot understand someone in a noisy factory? Regional or non-native accents might be unintelligible. The more variances, the more templates required. Beyond recognition lies the world of understanding, the domain of artificial intelligence, discussed later in this chapter. Figure 6.29 displays schematics for both recognition and synthesis. Spoken words are digitized into and analyzed and stored into ROM. To reverse the process the speech synthesizer converts the digitized ROM into analog audio output. The lower figure stores a library of speech templates. These templates are compared with the digitized words from above until a match is found. A coded command signal is then sent to the robot.

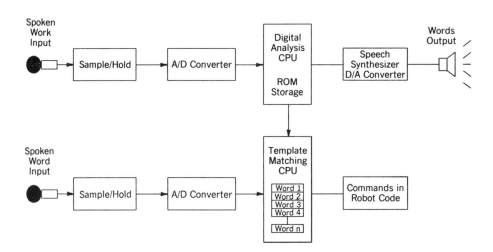

*Figure 6.29  Speech Recognition Circuitry*

# Control

## The Human Brain and Central Nervous System

Cogitative, memory and control functions are centered in the three pound brain. With over 10 billion nerve cells, it dominates the central nervous system and is housed in a rugged, sealed, generally spherical container, the skull. The blood flowing out of the brain is only a fraction of a degree warmer, an indication of its incredible efficiency. It consumes 20 Watts of power continuously; more power is consumed in concentration than in heavy exercise. The brain provides muscular coordination via the cerebellum and brainstem at an unconscious level. These areas of the brain communicate with the higher brain centers (primarily the cerebral cortex) to report on muscle status. Integration of senses occurs in the thalamus. Together with the cortex, the thalmus integrates input of sight, sound, touch, taste, and smell. Seventy percent of the human sensory cortex is devoted to vision.

Thought and memory reside in the cerebral cortex, which coordinates input from the senses and initiates voluntary actions. Emotions originate in the limbic system, which is buried deep within the brain. Fear, hunger, lust, rage, are just some of the emotions that begin there. The inner ear is a dedicated sensory mechanism, the semicircular canals providing constant positioning information to the brain for balance.

The spinal column houses and distributes the nervous system to the various organs, muscles and sensors. The brain branches to the cranial nerves and the spinal cord, a cylinder of nerve tissue that runs through the backbone for 18 inches. Nerves composed of thousands of fibers branch out from either side of the spinal column and spread to every corner of the body to provide signal response at speeds of over 200 miles an hour.

The brain and its multiple hierarchical functions have a robotic analogy in the computer, and the central nervous system in feedback servo loops. To begin, we will look at the more primitive background functions that fall under the category of control.

## Controllers

The function of controllers in early machine tools and robots was to store and execute simple programs that produce the sequence of motions required of a tool or endeffector at predetermined speeds. In doing so it relieved humans of the drudgery of repetitive machining or manipulations. Although now capable of more advanced interactive functions, controllers are still relatively primitive compared to humans. Control or servo systems often are defined using the biological model. Lower-level artificial intelligence function (i.e., brain stem functions) monitor and process information from sensors and regulate voluntary action. The three lower levels of control may be broken down into: (1) record/playback—point-to-point programming, (2) continuous path servo programming, and (3) structured programming. Higher levels include: (4) task level programming - the robot is interactive with its environment, and (5) human intelligence.

Control or servo systems may be traced back to the flyball governor invented by James Watt in 1788 to regulate the speed of steam engines. The faster the flyball spun, the less steam was admitted into the cylinder. It demonstrated one of the basic functions of control system regulation by mechanically regulating the throttle, instead of relying on a human operator.

Sir James Clerk Maxwell developed the servomechanism theory. His first servo mechanism paper "On Governors," published in 1868, formalized the basic concept that was already in practice. In 1893 Christopher M. Spencer filed his patent on a "Fixed-Head Type of Automatic Lathe for Making Metal Screws Automatically" (Figure 6.30). The key innovation was the blank cam cylinder, with its adjustable (dare I say programmable) curved plates for various jobs. In place of a skilled operator repetitiously creating identical parts, the operator's motions were built into the machine. A veritable money machine, the device gave birth to the tantalizing prospect of eliminating the human element from manufacturing.

Once these machine tools became electrically powered, the next step was to make their motors controllable by electrical and electronic means. Direct descendants of Spencer's "programmable" cams came in the 1940s when John T. Parsons of MIT led the development of a numerically controlled milling machine, guided by coded punch cards which machined the contour surfaces of helicopter blades (Scott 1985). One decade later the U.S. Air Force commissioned the MIT servomechanism lab to develop the numerical control machine language APT. Using the fledgling servo theory, motors on milling machines and lathes would be controlled by a programmable computer using punched tape. Machining operations too difficult or even impossible by humans could now be performed in less time, with increased profits. Servo technology was the key technology for the arm-like manipulators that would become known as robots.

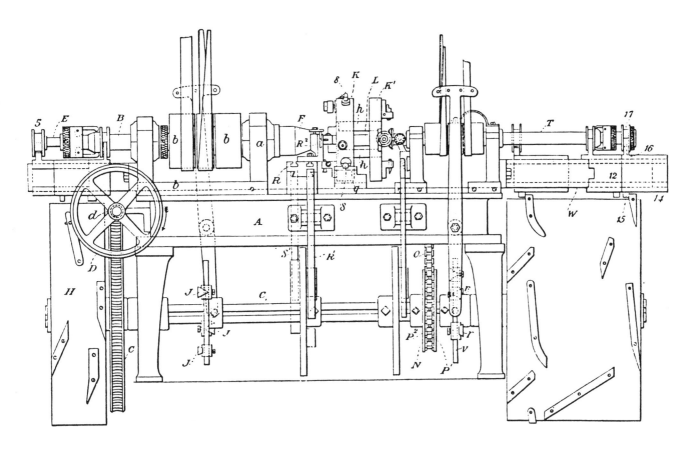

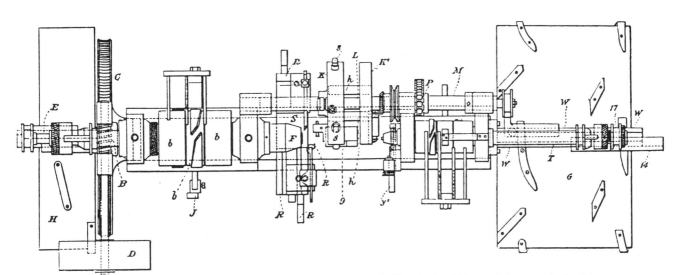

*Figure 6.30  Fixed-Head Type of Automatic Lathe*

## Planetbot Controller

Perhaps the first modern robot controller built was by the Planet Corporation for the Planetbot. The control system is reviewed in Chapter 2. Originally it was controlled by a mechanical analog computer resembling the margin setting mechanisms on old mechanical typewriters. This laborious process (not far removed from the Scribe's cam system) no doubt contributed to its lack of commercial success. It had two speeds: off and full speed. The mechanical arrangement was superseded by an electrical sequencer. One of the quirks of this machine that it would run wild in the morning because the hydraulic fluid temperature would be cooler than the previous afternoon (this was before the days of hydraulic fluid heaters for robots). From promotional material of its time comes the following description of the robot's operation:

> Each movement is connected mechanically to a multiturn or linear potentiometer. The amount of a movement can then be represented or measured by amount of resistance. After initiation of a movement, that movement will continue until its resistance matches a preset resistance in a plug board arrangement of the master control panel. When a match or dead band is made, the movement stops. Sequencing movements are accomplished by means of a variable time stepping switch.

> The master control panel contains four rows of jack sockets of 30 sockets each. Each row represents a movement; that is, main swing, tilt, retract, etc. In addition, there are two rows of potentiometers; one row of 10 black dials and one row of 12 red dials, which are used as reference or preset resistances to control the degree of any movement.

The now antique, premicroprocessor controller, more versatile and easier to program than the Scribe's stacked cams, nonetheless left much to be desired.

> A typical setup cycle would be as follows: The stepping switch is set to zero, if the first movement in the cycle is main swing, then a jack matching the color code of the first preset potentiometer is inserted in hole #1 of the swing movement jack socket row. The amount of swing is then controlled by the setting of the preset potentiometer. If the next movement is arm extend, a jack matching the color code of preset potentiometer #2 is inserted in hole #2 of the arm extend jack socket row. This setup may be continued until a total of 25 individual movements are predetermined.

Clearly, control technology needed to become much faster and more user-friendly if robots were to have a meaningful place on the production line. The early machine tools and robot controllers reproduced mechanical motions automatically, but the result was a memory storage device rather than a decision making machine capable of interacting with a changing environment (Hole, personal communication 1990).

### Unimation's Control Work

The first of a series of key patents by George Devol that would lay the foundation of level 1 record/playback programming was filed December 10, 1954 (Devol 1961, 1967). Disclosed was a track mounted, hydraulic robot arm (Figure 6.31). Position feedback information was provided by magnetic read heads connected to the moving portion of the arm and base, which scanned a magnetically coded surface composed of wafers of ferrous material. The breakthrough in this control system concept was the ability to control the rate of the arm's motion, slowing the arm before it stopped. This was a major advance over previous systems such as the Planetbot's controller. Position information was stored for retrieval on a magnetic drum. Programming was accomplished through a teach pendant. Although this particular development of encoding ferrous wafers was not commercialized, the basic concept of electronic programmable control was. Subsequently, the functions of encoder and memory device were divided between optical encoders at the joints and a continuously rotating drum

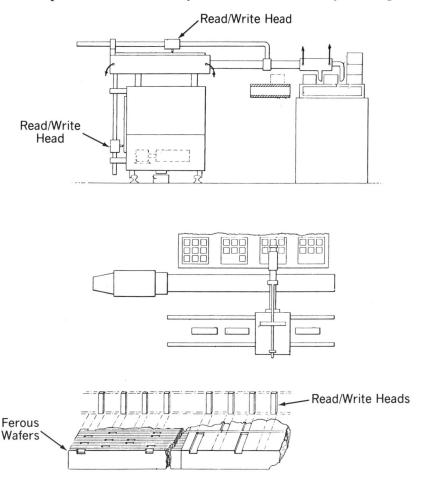

*Figure 6.31  Programmed Article Transfer*

memory in the controller which stored digitized position information (Dunne 1972). Quoting from this key patent:

> A programmed manipulator apparatus employs a continuously rotating memory drum in which may be stored signals representing a number of steps in a desired sequence of operations [Figure 6.32]. A portable teach control assembly [the teach pendant was a Unimation first] is employed during the teaching operation to establish a desired position of the manipulator arm which is then recorded on the memory drum.
>
> During repetitive work cycles, the recorded signals are used as command signals and are compared with encoder signals representing absolute position to move the manipulator arm to each set of positions in sequence. The manipulator arm may be moved over curved paths by employing artificial coincidence signals which are developed while large error signals still exist in the control axes. A common comparator and digital-to-analog converter is employed for all of the controlled axes which provides coincidence signals representing different magnitudes of error.

The Unimate controller resembled the Planetbot controller with its rows of dials to set speeds for each movement. Later, computer interfaces were eventually added to increase versatility.

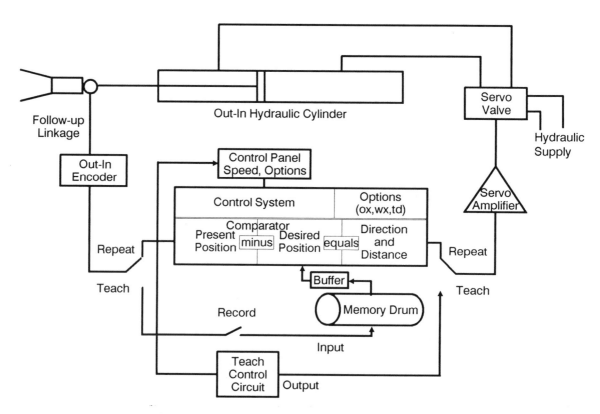

*Figure 6.32  Early Unimation Control Schematic*

## Cincinnati Milacron's Control Work

Beginning with "Method and Apparatus for Controlling an Automation Along a Predetermined Path" filed May 31, 1973 (Hohn 1975), Richard Hohn and other Milacron engineers filed a series of basic patents on the continuous path servo control (level 2) of revolute arms (Figures 6.33 and 6.34). Improvements in control were required for the more articulate, large work volume arms that Milacron created. Unimation's arms were of a simpler spherical work volume and were matched to fit their more primitive controller. Microprocessor based computers were essential to perform the interpolation required to replace the original spherical coordinate system with the more user-friendly rectangular coordinate control system. Richard Hohn's patent led to a simplification of programming of a robot's path between two endpoints which became known as the tool center point (TCP) concept. Coordinate conversion, storing and interpolating position data in rectangular coordinates, was accomplished. The interpolation provides constant velocity and maintains the endeffector at a prescribed attitude along a straight line.

Figure 6.34 is a block diagram of the Milacron controller. The following is adopted from the patent: From an input (left group of blocks) signals are generated representing the predetermined velocity and rectangular coordinate values of end points of the predetermined path; the program operates on a fixed time basis to iteratively calculate a number of incremental displacements. Coordinate values defining each of the increments are calculated, and these coordinate values are transformed into corresponding generalized coordinate values defined by the geometry of the arm. A servo mechanism driver circuit (the group of blocks to the lower right) responds to output signals representing changes in the generalized coordinate values along each incremental displacement for causing actuators on the arm to move proportionally. In operation the user describes a beginning and endpoint in a rectangular coordinate system, and the controller automatically translates this command to the arm's spherical coordinates and the joint angles necessary to achieve the endpoint.

This automatic translation from one coordinate system to the other is a basic, lower-level function for anthrobots. As with humans, control of anthrobot limbs requires unconscious transformations to grasp moving objects, avoid collisions, manipulate things, and walk.

In a subsequent patent filed in July 16, 1974 (Corwin 1975) this concept was further developed so that an operator could manually program a robot through a teach box, with the robot's computer automatically translating the rectangular coordinates to the revolute arm's spherical coordinates (Figure 6.35). The operator "instructs" the robot to move in a series of points along a path, and the robot's computer connects the movements together to form the entire cycle of operation. Prior to this

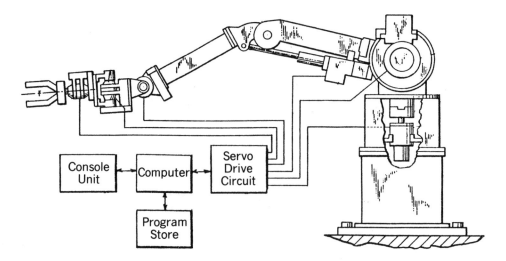

*Figure 6.33  Early Milacron Control Schematic*

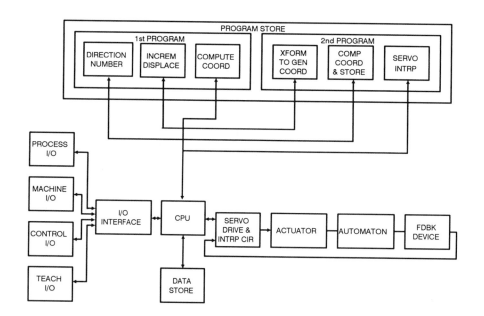

*Figure 6.34 Early Milacron Control Block Diagram*

point-to-point concept, each axis of the robot was discreetly programmed by a teach box. Also, each axis moved at a fixed rate, with the axis having the shortest displacement completing its motion first. Consequently, an irregular path is created that makes it difficult to predict the arm's motion around obstacles.

Additional Milacron control patents would build on this base, further developing options and basic controller architecture. One example is toolpoint tracking of moving workpieces (Hohn 1977). This allows the robot to synchronize with a workpiece on a conveyor belt automatically modifying its position. Milacron's robot division was purchased by ABB in 1991.

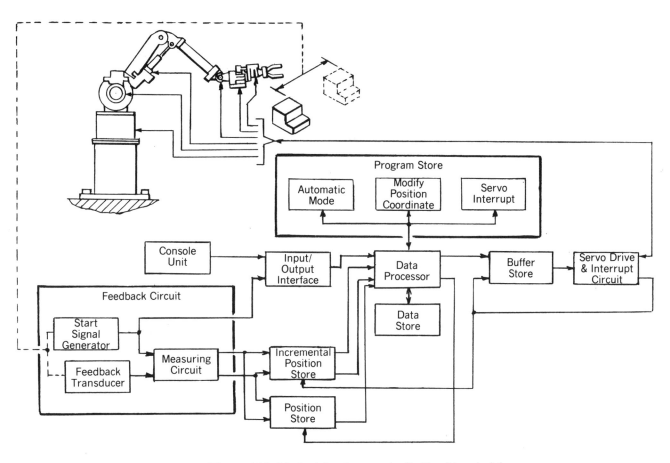

*Figure 6.35 Method for Automatically Tracking Articles*

## Robotics Research

The state-of-the-art in structured programming control technology (level 3) is embodied in the Robotics Research arms and controller. Robotics Research developed the first commercial manipulator design to successfully employ harmonic drive/electric motor actuators as direct joint drives (Karlen 1990). This advance was made practical because of an innovative, torque loop, servo control system.

> Semiconductor strain gauges mounted in the actuator are used to measure and regulate harmonic drive output torque, providing control stability over a wide range of inertial conditions. This feature also affords means to achieve joint torque control, with a very high bandwidth and thus toolpoint force control. It provides feedback for teleoperated systems with bilateral force reflection, and makes an otherwise high impedance actuator mechanism have electrically controllable and programmable compliance (personal communication, Karlen 1992).

These arms are coupled to powerful control algorithms which make it possible to control stiffness. This is necessary with normally compliant, high friction, and transmission error prone harmonic drives. The open architecture, NASREM based motion control system allows the user to operate the manipulator in either robotic or teleoperated modes using Cartesian toolpoint coordinates or by using joint position, velocity, and torque commands. NASREM refers to the "NASA/NBS Standard Reference Model of Telerobot Control System Architecture" developed by the National Institute of Standards and Technology for NASA (Albus 1987). This control technology also allows lockup of selected joints. Joint lockup is important with a seven degrees-of-freedom arm; it prevents a situation akin to wrestling a boa constrictor when all axes are free. The programmer can also select a weightless operating mode for the arm, decoupling it from gravity and from the drivetrain's mechanical friction. This type of control system is a far cry from the complicated and crude counterweighted arms of GRI and GMF, in which low friction and springs are required. Power to the arm can simply be reduced for operator safety during walk through programming.

Beyond the relatively simple applications of walk through programming and teleoperation, torque sensing in the joint makes other control modes possible: Reflexive Collision Avoidance, in which obstacle avoidance and joint travel limits are provided by mathematical "force fields" created around each obstacle point; Suspension emulation, in which abrupt endeffector motion that could result in high torques is reduced by dissipating the manipulator kinematic energy over longer time periods; Impedance Control, which may be implemented to vary joint stiffness. Torque sensing also permits mechanical advantage and positioning resolution management. This technique is applied to singularity avoidance and capitalizes on the mechanical advantage often present when a near singularity condition exists. Also, torque management and redistribution can be provided on the individual joint level by selecting spring stiffness and dampening of inertias. Favoring an overloaded joint is one example of torque management. Finally, torque sensing allows pose optimization. As in the human body, this control technique is designed to increase efficiency through preselection of poses that provide the greatest mechanical advantage to the given task.

As the complexity of robot applications grew, so did the need for specialized computer languages. NASREM, VAL, RAIL, and KAREL are but some that now exist. Increasingly, offline programming, spurred on by the development of high powered graphics, has made programming more efficient and user-friendly.

Like its human counterpart, the robot moves its endeffector as a sum of several coordinated motions without "thinking" about the details, each joint's angle, speed, and sequence being determined in a background mode. The Robotics Research patents (Karlen 1990, 1992) explicated control techniques (level 3) for modifying the command signals that represent the predetermined positions of a machine element relative to a workpiece (Figure 6.36). For example, if a fixed workpiece is moved or a conveyer speed is changed, the existing program may be modified to

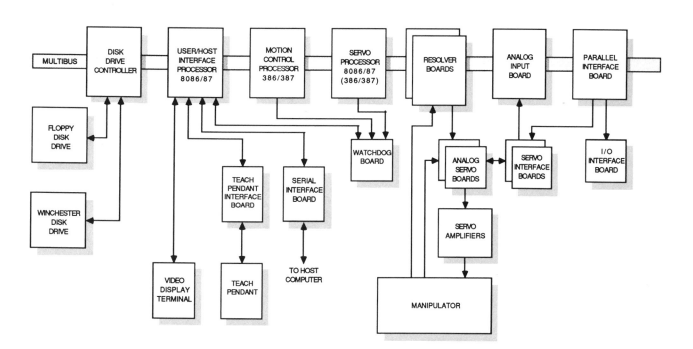

*Figure 6.36  Robotics Research Control Architecture*

accommodate the changed workpiece position. This ability to interact with its environment represents even greater sophistication and is essential to anthrobotic anatomy. Such control systems, if fully developed, will enable adaptation to changing, free-form environments such as those that humans inhabit.

In future anthrobots, control systems will monitor and regulate homeostatic functions (orientation, temperature, energy levels) in a background mode with artificial intelligence monitoring both the background and foreground, focusing attention as needed. Automatic functions will include automatic shifting from one coordinate system to another. These future control systems must be massively parallel to integrate this large amount of internal and external information and regulate outputs; this implies neural networks and machines that learn.

## Artificial Intelligence

We have looked at controllers which are principally record and playback devices. To reach the task level goal in artificial intelligence (level 4), these devices must be surpassed.

For task planning yet unobtained, the robot actions are described by their effects on objects in the task environment, not by specifying manipulator motions necessary to complete an action. For example: Picking up a cup requires gross motion planning. Next, the grasp planner reflects on where to grasp an object to avoid collisions when grasping, moving and resetting. Path is also important to prevent spillage. Fine motion planning is heavily dependent upon sensor feedback. The hand might touch the cup before grasping to give it the benefit of a "biased search" because once the hand has a real world reference it makes grasping much easier. Computers were first developed for performing mathematical calculations which were tedious and time consuming. Later, navigation and military applications created whole new classes of problems unsolvable by humans. As basic computer power grew and input/output devices advanced, new applications were developed for diverse nonscientific applications.

Two basic approaches are used by Artificial Intelligence engineers. The top-down approach attempts to gain design insight from studying the organic brain. Unfortunately progress in this area is slow, for if you cannot understand the human brain you cannot model it. In the bottom-up approach engineers seek to design artificially intelligent computers based on their present knowledge of computers and software. Perhaps in the end a synthesis of these approaches will be required, for even if the bottom-up approach appears to work we cannot know that the processes underlying functionality will be the same as ours unless we know how humans think.

As marvelous as it sounds, human intelligence (level 5) may not be desirable for anthrobots. Who wants rebellious slaves or something so versatile that it may be purchased only by large organizations? Large markets are forecast for robots that can contentedly function as domestic servants (Engelberger 1989). For such tasks we do not need a mechanical malcontent who complains about being overqualified to park spaceships. Although the ultimate technical challenge for anthroboticists may be the creation of a robot that in all ways simulates human capability, in truth such an outcome is economically unnecessary and probably morally repugnant. In contrast, making humans is simple, fun, and an overabundance already exists.

In the following a review of computer hardware and the pioneers who built them, the technical benchmark is human-like functionality. The desirability of achieving such a standard, while a worthy subject of argument, is material for another essay. Level 4 task level programming provides the anthrobot with the functional capabilities to perform most human operations.

## Analytical Engines

Charles Babbage (1791-1871) English mathematician, inventor and reformer, designed mechanical computers that anticipated almost all of the basic functions of the modern computer. Building on the work of Pascal and Leibniz, who built a calculator to multiply, add, divide, and extract square roots in 1694, Babbage's most fascinating contribution, though never completed, was his Analytical Engines (Figure 6.37). As his biographer Anthony Hyman says in *Charles Babbage Pioneer of the Computer* (Hyman 1982).

> Common to Analytical Engines and to computer are: the separate store for holding numbers and mill for working them; a distinct control or operations section; punch card input/output system;...(page 164).

> He [Babbage] planned a whole range of peripherals, including a card reader and punch, line printer, and curve plotter. These could work locally or remotely (in the next room). He could effect several operations simultaneously, including transfer of numbers between the store and mill while operations were being performed in the mill, thus taking one or two steps toward the modern technique of pipelining (page 209).

*Figure 6.37  Analytical Engine*

Array processing, the technique of using several processors to solve large problems, the IF and DO statements, and conditional branching were also featured. This work was initially funded by the British government but they withdrew support. The machine was handicapped by decimal notation, making the device enormously more complicated than the binary coding adopted a hundred years later. Had funding been continued, the spinoff advancements in machine tool technology alone would have been enormous. Recently, one of his early machines "Difference Engine Number 2" has been constructed and is operational. Babbage's level of creative insight was not equaled until the work of Alan Turing.

## Turing Machine

Alan Turing (1912-1954) was one of the true pioneers in conception and development of the modern computer (Hodges 1983). He foresaw both computer communication via telephone and programming decades before their implementation. He wrote a seminal computer science paper, "On Computable Numbers" at the age of 25. He also developed the Colossus (an electronic cryptoanalytic machine), an early dedicated function computer, to crack the German Enigma code machines during World War II (one version built in 1944 used 2,400 vacuum tubes). Input was through an electromechanical punched card system originally developed by Herman Hollerith (1860-1929) (perhaps inspired by Babbage). It was one of the first dedicated, stored program computers. The Turing machine, on which all digital computers are based, started as an exercise in designing a super typewriter. To simplify things, the paper would be a tape of infinite length capable of one line of writing. In addition to writing, the machine could read or scan and erase. The goal was to design an automatic machine capable of reading a mathematical assertion, and eventually writing a verdict as to whether it was provable or not.

The Turing machine works independently, reading and writing automatically as its read/write head moves back and forth according to its design. Direction and operation of the head is controlled by its state and the character it has read. The machine's design dictates how it processes the following operations.

Write a new symbol in the blank, erase or leave an existing symbol.

Maintain present configuration or reconfigure to a different configuration.

Move right, left, or maintain present position.

Shown is a depiction of a Turing Machine and **Table 6.2** defines the machine as an adding machine.

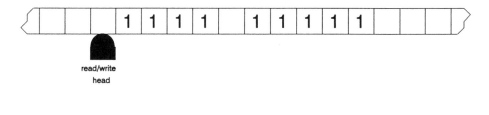

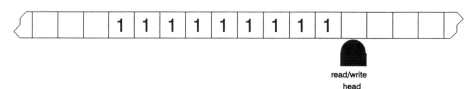

The table of behavior is the "program." The head "recognizes" whether the box is empty or not and "decides" its next move. Thus, it features two key elements that would lead to the modern computer: recognition and decision making.

*Table 6.2  Scanning Machine Table*

| Configuration | Symbol read | |
|---|---|---|
| | blank | 1 |
| *1* | *shift right; config. 1* | *shift right; config. 2* |
| *2* | *write '1' shift right; config. 3* | *shift right; config. 2* |
| *3* | *shift left; config. 4* | *shift right; config. 3* |
| *4* | *no shift; config. 4* | *erase; no shift; config. 4* |

The linear process that makes Turing machines easily programmable limits them to certain types of problems. Although the human brain was the model throughout Turing's work, the brain works in a parallel process with several interactive tapes, not one, in operation. After the War, Turing went on to build the Automatic Computing Engine (ACE) (Figure 6.38) in 1950, implementing the concept he pioneered. The first commercial machine using these principles was built by the Ferranti Company in 1951: the Mark 1.

Turing is also well known for the "imitation game," or Turing Test, published in 1950 (Hodges 1983), a method of determining human functionality of computers. When communicating with a computer, if the human cannot distinguish the computer from a human, then for all functional purposes the computer is "human." This test is also the basis of the laws of anthrobotics as stated at the beginning of this chapter. Turing predicted that by the year 2000 an interrogator would only be 70 percent accurate in identifying the subject after five minutes of questioning. In 1991 a limited Turing Test at the Boston, Massachusetts, Computer Museum was run, fooling five out of the ten judges. The winning program, "PC Therapist," by Thinking Software, Inc. of Woodside, Queens, New York, engaged the judges in a whimsical conversation. Strangely, two human confederates were guessed to be computers (Markoff 1991, 1993). Whether "thinking" is going on inside the computer some would argue is irrelevant. Function is the primary concern for the anthrobot. However, Artificial Intelligence pioneers will not rest until even this is accomplished.

*Figure 6.38 Automatic Computing Engine (ACE) 1950*

## Logic

Anthrobots need reasoning techniques to accommodate unstructured environments and situations. Developed by the ancient Greeks, logic is a formal method of reasoning. Building on a foundation of known true facts, new facts can be inferred that are certain to be true. Propositions, statements that are true or false, are linked with connectives (i.e., And, Or, Not, etc.). Deductive procedures known as inferences are used to determine whether a series of propositions is true or false. Thus, proposition 1 and 2 are true, the result 3 is true. Another example of an inference is:

Management is incompetent.

Your boss is a manager.

Your boss is incompetent.

Branching out from this simple root is predicate logic. It permits inclusion of assertions about items in statements and variables. Thus the inference example would read as: Incompetent (manager, your boss). Logic has evolved into a symbolic language all its own that lends itself to processing by a computer.

The resolution method uses the procedures of evaluating a theorem based on a given set of axioms. Operating under the premise that the conclusions are false is the premise adds up to an impossibility. Thus the original conclusion is proven true or false.

Fuzzy logic was invented by Professor Lofti Zadeh at the University of California at Berkeley in the mid-1960s. Unlike grading elements in black and white terms, fuzzy logic allows for shades of gray. This allows for more realistic representations of a given situation. Also flexibility makes the task of describing situations simpler than using mathematical equations (Miller 1993).

Frank Severance, Assistant Professor at Western Michigan University describes it thus. "What fuzzy really comes down to is that it is an intuitive way of controlling systems as opposed to traditional control systems which are very mathematical. With traditional control systems you cannot start until you know the mathematical model of what you are trying to control. With a fuzzy logic controller, you assume that you are working with a non-linear system that you frankly can't describe very well. There-in lies the beauty of fuzzy logic—the systems are simple, yet they work."

Logic is the core on which all computing systems are based. Expanding on the general concepts of logic has led to expert and other artificial intelligence systems.

# Computers

## ABC

Iowa State Professor John Vincent Atanasoff (1903- ) conceived and built, with the assistance of Clifford Berry, the first stored program digital computer in 1939 (Figure 6.39). Known as the Atanasoff-Berry Computer (ABC), it was designed to solve 29 linear algebraic equations with 29 unknowns. Outlined in a 1940 manuscript "Computing Machines for the Solution of Large Systems of Linear Algebraic Equations," it featured several firsts: all computing was electronic (with the exception of the clock); vacuum tubes were used for the first time, as well as a binary number system, logic systems, computation in a serial manner, drum memory using capacitors, and memory regeneration. The machine employed over 300 vacuum tubes and used punch cards for input. This desk-sized computer became operational in 1941, although not all the bugs were worked out (Mollenhoff 1988).

By employing a binary encoding system, rather than decimal as Babbage's had, Atanasoff and Berry's machine could make errors but still be very accurate. By making their machine electronic rather than mechanical they gained tremendous speed in calculation. Atanasoff shared his knowledge with John Mauchly, thus influencing the design of ENIAC, generally recognized as the first stored program electronic computer.

*Figure 6.39  Atanasoff Berry Computer*

## ENIAC

Electronic Numerical Integrator and Computer (ENIAC) was built at the University of Pennsylvania's Moore School of Electrical Engineering. Began in 1943, it was designed for solving Army Ordnance ballistic problems (Figure 6.40). J. Presper Eckert and John W. Mauchly were its principal architects. Weighing in at 30 tons, its forty panels filled a room 30 feet by 50 feet (9 x 15 meters) and was equipped with approximately 17,000 vacuum tubes, 1500 electromechanical relays, 70,000 resistors, 10,000 capacitors and 6,000 switches, all of which were discrete components. When powered-up, neighborhood lights dimmed.   A 60 second missile trajectory calculation that took 20 hours and 15 minutes on an analog differential analyzer took only 30 seconds on ENIAC (Clark 1985). However, vast amounts of wiring had to be accomplished before ENIAC could begin to solve a problem.  If a tube failed it could take eight hours to find the fault.  ENIAC's operational or up-time was one hour.

*Figure 6.40  Electronic Numerical Integrator and Computer*

## Von Neumann Architecture

John Von Neumann (1903-1957) one of the 20th century's most important mathematicians, obtained his Ph.D. at 22. He had the political savvy along with the connections and influence to get the early computers built. Von Neumann described his machine in symbolic terms, using the analogy of human neurons. Von Neumann wanted to build a computer version of a brain, but due to practical necessity had to make the concept serial, not parallel, in operation. By ignoring the minute engineering details, he was able to arrive at a clearer understanding of basic computer concepts. He was, in effect, painting in broader strokes than his predecessors, and produced a large canvas outlining the stored program computer as we know it today. The architecture outlined in "First Draft of Report on the EDVAC" (1945) organized the computer system into five units that are still familiar. William Aspray in his definitive book *John Von Neumann and the Origins of Modern Computing* (Aspray 1990) summarizes the contents as follows:

> A central arithmetic unit to carry out the four basic functions such as roots, logarithms, trigonometric functions, and their inverses; a central control unit to control the proper sequencing of operations and make the individual units act together to carry out the specific task programmed into the system; a memory unit to sort both numerical data (initial and boundary values, constant values, tables of fixed functions) and numerically coded instructions; an input unit to transfer information from the outside recording medium to the central processing units (the arithmetic unit, control, and memory); and an output unit to transfer information from the central processing units to the outside recording medium. (page 39)

The Electronic Discrete Variable Automatic Computer (EDVAC) was a first generation machine built at the Institute for Advanced Study (IAS) at the University of Pennsylvania during 1946-1951 (Figure 6.41) and became known as the IAS computer. An engineering tool, this programmable calculator was Turing in hardware. Von Neumann's second major paper, "Preliminary Discussion of Logical Design of an Electronic Computing Instrument" laid out the basic logical design known as Von Neumann architecture (Figure 6.42), a general purpose computer composed of arithmetic, storage control input and output units (Aleksander & Burnett 1987). Input was by IBM punched card and output by modified teletype printer. Circuit elements included 3,000 tubes that consumed 28 KW of power. It occupied 36 cubic feet and weighed approximately 1,000 Lb (454.5 Kg). Clocktime was 17 microseconds, internal storage was 1K to 4K of memory, and word length was 44 bits. More advanced, commercial first generation machines followed, such as the Univac I, which employed symbolic languages rather than raw machine code.

*Figure 6.41  IAS Computer*

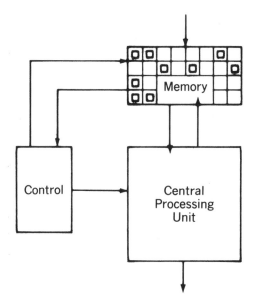

*Figure 6.42  Model of Von Neumann's Architecture*

Also a pioneer in artificial intelligence, Von Neumann articulated the Minimax theorem, a form of data acquisition fundamental to game theory. In this algorithm, the machine will always try to maximize its own score when it makes a move, whereas its opponent will always try to minimize the machine's scoring possibilities on his turn. The machine is able to look ahead several moves, starting at a single node and branching out from this in a decision tree (Figure 6.43). As the depth of the decision tree increases, so does its quality of moves. Searching at the final level, the machine looks for the highest score, expecting that its opponent will try to minimize the machine's score. The machine "backs up" this selection to the level above (which would be the opponent's turn), choosing the lowest score. This process continues until the path along the best score is traced, thus representing the best move.

One of the major problems with the Minimax algorithm, however, is searching a decision tree such as this can be very expensive, both in CPU and realtime. To "look ahead" requires time, which in some realtime situations may not be available. One method of improving this search time is by the use of "alpha-beta" pruning. In this method, the decision tree is inspected and complete branches are eliminated from further search on the basis that the path would not be followed using the current heuristics. This can speed the search up considerably, allowing more levels to be considered.

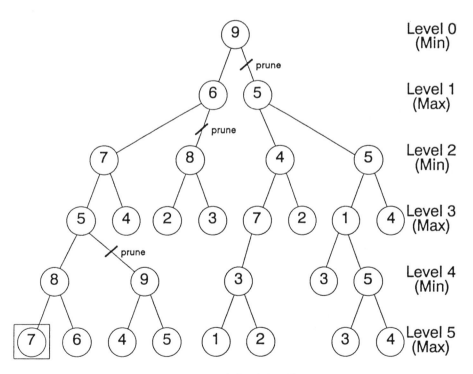

*Figure 6.43 Decision Tree*

**Second Generation Computers (1959-1964)**

Following in the wake of machines like IAS were machines such as the Univac 1107 (Figure 6.44), which was introduced in 1962. It featured the same basic architecture but used the major advancement of transistor circuits, which greatly reduced weight, size and power consumption. Faster magnetic cores replaced magnetic drums for internal memory. Clocktime was 4 microseconds, internal storage was 32K to 65K of memory, and word length was 36 bits. Punched cards were replaced by magnetic tape for input, which was also used for storage. This period also saw the birth of the first high level programming languages such as FORTRAN (FORmula-TRANslator), intended for scientific applications, and COBOL (COmmon Business Oriented Language), for business purposes. Also, advances in memory devices were made. They were adopted as data processing engines. Computer time was so expensive that operations were performed in batches for cost effectiveness, not user convenience.

*Figure 6.44  Univac 1107*

### Third Generation Computers (1964-1975)

The next level of computer hardware was introduced in 1965 and is exemplified by the Control Data Corporation 6600 (Figure 6.45). Its principal architect was Seymour Cray. Key features included large scale integrated circuits (LSI), further reducing size, decreasing power and increasing speed. Also, LSI circuits lend themselves to cheaper mass production, thus reducing cost and increasing availability. Clocktime was 1 microsecond, internal storage was 32K to 131K of memory, and word length was 60 bits. Significant advancements were made in application of high-level programming languages and solid state internal memory. These computers supported multiple users because the computer operating system was now capable of diverting its attention to several tasks.

*Figure 6.45  Control Data Corporation 6600*

## Fourth Generation Computers (1975-?)

Further expansion of large scale integration to very large scale integration (VLSI) gave rise to the personal computer and super computers. The CRAY 2 super computer is an example of a state of the art fourth generation computer (Figure 6.46). It contains a vector/scaler processor, so named because it contains special circuits to calculate large arrays of numbers in long lists or vectors. One instruction carries out all needed calculations on the entire list of numbers. The term scaler refers to the modular Central Processor Unit (CPU) and may be scaled up or down depending on need. Features include a large common random access memory (256 million 64 bit words), multiple processors, and liquid immersion cooling. The latter is one of the most interesting physical features of the machine. To achieve high density and thus high speed (clock cycle time is 4.1 nanoseconds), the logic and memory circuits are contained in 8-layer, three-dimensional modules. To drain away the considerable heat output of such a configuration they are immersed in constantly circulating fluorocarbon fluid. The three-dimensional design of the circuitry emulates the design of the human brain. One of the problems in creating programs that function like human thought processes is the scarcity and cost of this type of hardware. Many believe it is this type of computer that is essential for neural networks to take off.

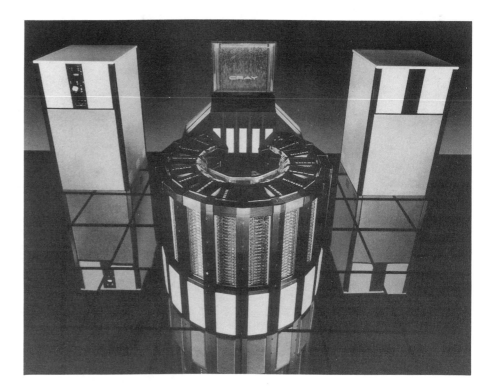

*Figure 6.46 Cray 2 Super Computer*

## Fifth Generation Computers (? and Beyond)

These will be massively parallel processing machines (MPP) operating in the Teraflop range (trillions of instructions per second). The Teraflop is considered to be a milestone beyond which whole new classes of problems, known as "Grand Challenges," are expected to be solved, including natural language input, vision, and other artificial intelligence goals. A Teraflop machine is predicted by Cray Computers in 1997. Two basic approaches are contenders: Thinking Machines Corporation links tens of thousands of conventional processors in parallel to work simultaneously on different streams of instructions. The traditional approach pioneered by Cray Research is to link expensive custom made vector processors. Figure 6.47 is a graphical representation of the two approaches.

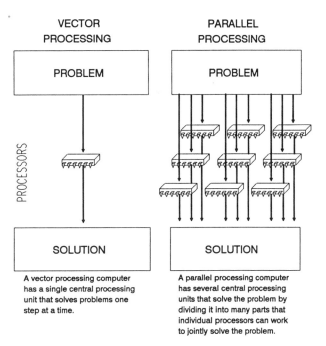

*Figure 6.47  Vector vs. Parallel Computer Processing*

The power of at least a fourth generation is what is needed for anthrobots to become a reality. If size, power consumption, and cost can continue to be reduced at the current rate (50 percent reduction every three years) (Tesler 1991) perhaps by the year 2000 a supercomputer may be built that could provide level 4 control for an anthrobot. A reduction of cost, size, and power reduction may have to be as large as the difference between the EDVAC and CRAY-2, thus prompting the designation "sixth generation."

Even the CRAY-2 cannot perform most of the perceptual functions we perform with little effort. With a 16' footprint, a weight of 5500 Lbs. (2500 Kg) and an elaborate coolant system, this state of the art machine could hardly be packaged into a human scale anthrobot. Current input technology also needs an improvement factor as large as the leap from punched cards to current speech recognition systems, from keyboard to full natural language input.

## Expert Systems

Expert systems (Figure 6.48) provide a distillation of a particular area of knowledge or expertise. More than a reference tool, they comprise a flexible interactive system that strives to replicate human problem solving methods. The programs are compartmentalized into rules or IF-THEN statement sections, and a database of facts. Harking back to Babbage's terminology, expert systems are inference engines. Like neural networks, the expert system "weighs" several factors before producing an action or opinion.

In Joseph Engelberger's excellent, practical book, *Robotics in Service*, (pages 105-106) he describes an expert system for a robot watchman.

> The human trainer feeds the system with all the rules of deportment essential to the robot's workspace. Facts regarding that workspace are fed into the database. When the robot receives a command the inference engine uses the appropriate rules and the facts at hand to give the robot detailed instructions. The robot as well as the trainer can enrich the rule based system by reporting on how the task proceeds, on missing facts and on conflicting rules. Interactive, the system gets smarter until the human is comfortable with the entire range of tasks in the robot's workspace.

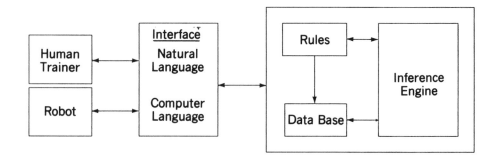

*Figure 6.48 Expert or Rule Based System*

The following might be a sequence of actions taken by a household robot under rule based interpretations of a unique situation.

If heat sensor alarm, then there is a fire.

If a fire, then awakens all humans.

If humans awake, then escort them out.

If humans out, call fire department.

If above actions completed, seek fire.

If fire found, extinguish it.

If fire extinguished, report to humans.

The difficulty with expert systems is their reliance on algorithms written by humans. To accommodate every different situation an algorithm must be written, which can produce an impossibly long list of rules. The name "combinatorial explosion" has been given to this phenomenon which occurs when machines interact with our unstructured environment. Originally the TRC Helpmate had a few hundred rules to help it cope with the hospital environment. This soon blossomed into a few thousand (Feder 1992). This is apparent in the current efforts to make self-piloted vehicles whose "brains" fill large recreational vehicles, yet travel at very modest speed. Expert systems can only be as smart as their experts; garbage in means garbage out.

### Semantic Networks

Another useful AI technique is semantic networks, which are a more efficient method of representing a knowledge base than logical formulas. They use linking tools, such as IS-A, LOST-PART, or MOVED-TO rather than IF-THEN. Figure 6.49 displays a model semantic network. Objects or concepts are connected by links that describe the relationship. Nodes (the circles) may also contain events, situations, and concepts that create a network of knowledge for the anthrobot. This is a powerful way of representing knowledge, as it provides for a mechanism called "inheritance." For example, when we say that "Thomas Edison" IS-A "Inventor," it allows us to assume that "Thomas Edison" will have most or all of the characteristics of an "Inventor." By definition, he is a human who creates new devices and increases our knowledge. Instead of deriving each of these characteristics using logic (Inventors are humans. Thomas Edison is an inventor. Thomas Edison is human. Inventors file patents....and so forth). The semantic definition of "Inventor" allows the network to infer that Thomas Edison partakes of inventive characteristics. By using semantic networks, it allows us to represent an object's characteristics in a hierarchical manner, thus simplifying the network.

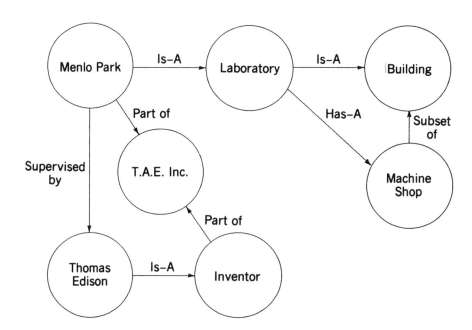

*Figure 6.49  Semantic Network*

## Frames

Yet another way of representing knowledge is a technique called "frames" created by MIT's Marvin Minsky. In frames, stereotyped objects, events or situations are stored so the system can extrapolate consistent data and apply it to similar, yet new, applications. Slots in the frames provide relevant details. For example, the factual frame for mowing the front yard would include: factual information such as boundaries, grades and obstacles. Heuristics frames would help to deal with unexpected situations, such as children or animals in the way. The combination of physical knowledge and heuristics recreates the human operator's skill. With this knowledge and reasoning base, the anthrobot, aided by sensors, would be able to extrapolate how to mow the back yard with equal elan.

## Neural Networks

Neural Networks began with Warren McCulloch and Walter Pitts, who first modeled human neurons in 1943. Norbert Wiener, father of cybernetics, saw the combination of logic and network structure as key to understanding human as well as machine thinking. This was incorporated in his seminal book *Cybernetics* (Wiener 1953) which linked biological to machine systems. Both Alan Turing and Von Neumann were well aware that the algorithm based machines were limited and discussed neural networks. The structure of the machine, not the human, algorithm would perform the computation. As noted earlier in this chapter, the vast majority of computers are based on the Von Neumann architecture, which was designed with practical implementation in mind. Computer architecture was constrained by the technology of the era, like the early industrial robots, whose mechanical design was limited by primitive controls and vice versa. Although the computer has dramatically improved in performance, the basic design is the same. Algorithms are used in which there is an inherent computational limit. These machines can only perform operations for which algorithms have been written. They do not learn; they only do what they are told (Aleksander 1991).

In contrast, neural networks are modeled after the human brain and require less rigid programing techniques than conventional digital computers. Emulating the human neuron's parallel operations—which individually are slower than computer circuits, but compensate by parallel operation (Figure 6.50)—they "learn," like humans, "by example" through learning algorithms. Emulating the human's ability to come to conclusions independent of a given set of rules, the system recognizes and generates its own rules by means of information feedback techniques. Training, unlike "programming," provides many examples of the correct answer per set of questions. A neural net is trained by repeatedly being "shown" a set of inputs and corresponding output. The more times it is "shown" the input/output matrices, the better it will "recognize" the input when shown and queried. Learning algorithms are more flexible than expert systems and are more adaptable to change, because they imitate the organic interconnection of all neural logic elements. The intelligence of these interconnections is described as weight functions. How important any one interconnect is to a problem determines its weight. Thus every decision is the result of a complex interplay of multiple factors, not isolated decisions.

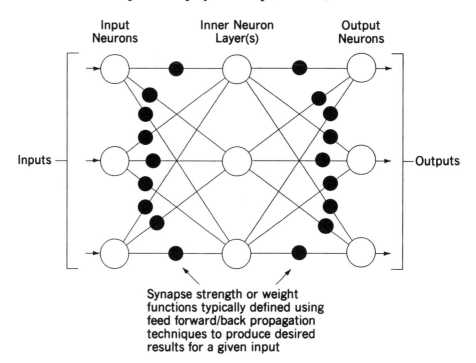

Synapse strength or weight functions typically defined using feed forward/back propagation techniques to produce desired results for a given input

*Figure 6.50 Simplified Neural Network*

In contrast to rule based systems, you cannot determine how the answer was obtained in a neural network. Like humans, it may find the right answer for the wrong reasons and vice versa. This is why a neural net works so like a human and why it is so potentially disturbing; we can no longer determine what it is thinking, we can only look at the results. Like a human brain, neural networks are faster than computers at approximations. New hardware concepts such as optical devices, rather than sequential digital computers, seem to be the way of the future. Their three-dimensional nature will allow structured cross talk between neurons to be achieved more readily (Compton 1991).

The planar nature of circuit boards does not lend itself to compact interconnection between circuits. To achieve greater parallelism, a three-

## CYBERNETICS

Published in 1948, *Cybernetics* was written by Norbert Wiener (1894-1964) during his tenure as a professor at MIT. A child prodigy, Wiener was one of the first to sketch the basic architecture of the modern computer. Cybernetics, derived from the Greek "steersman," created a sensation by establishing the relationship of biological systems to the nascent control and computer science field. Whereas Tesla's modeling of the human as a machine was limited by the control technology of his time, Wiener went into greater detail. Utilizing the corpus of knowledge available in the areas of the brain and nervous system he linked this to the fledgling computer science field.

Wiener drew the parallel to the neural network computer concept of McCulloch and Pitts. He saw computer concepts as useful in modeling the nervous system and vice versa. One of the first to describe proprioceptive functions, he found direct analogies of servo system "hunting" in humans known as "purpose tremor". A state of "ataxia" describes a proprioceptive function breakdown.

An early computer evolutionist, he stated: "In this program there was a gradual progress from the mechanical assembly to the electrical assembly, from the scale of ten to the scale of two, from the mechanical relay to the electronic relay, from humanly directed operation to automatically directed operation;..." Wiener foresaw the negative potential of automation on labor and tried to warn trade union officials. Fifty years ahead of his time, Cybernetics did not grow at the rate he predicted. It does stand however as a milestone in the connection of biological to mechanical systems.

dimensional hardware structure that lends itself to a high level of interconnection like the human brain is indicated. Cray Computers is leading the way by stacking chips on top of chips for greater speed. This may be the solution to the anthrobot vision problem.

Machine vision as currently practiced is far below that required for anthrobots. Fuzzy logic, cousin to neural networks, has been applied to video and other consumer electronic devices. Like neural networks, fuzzy logic learns from a history of inputs that train the circuits to produce a desired output. The camera learns the correct focus by repeated examples from its operator. Fuzzy logic techniques could be developed for anthrobots to learn walking. Given repeated visual and physical examples by their programmer "parents" the infant anthrobot would learn by example, or trial and error, just like its human counterpart.

## Self-Diagnostics

Anthrobots will need some built-in-test/diagnostics to simplify servicing, also known as "health and status monitoring"; inspiration for this comes from multilayer built-in-test (BIT) equipment and diagnostics in electronic equipment, including equipment as mundane as copying machines. The ability of the anthrobot to know how it "feels" is crucial for its performance.

Computers will also have to be compact and modular. Military computers, particularly avionics, are leading the way at present. Anthrobots must approach Mil-Spec ruggedness, as there's little point in building an anthrobot whose body is shockproof and waterproof, but whose chips come loose or fail when the anthrobot stumbles. Here again, the military's operate through standards may apply. This will result in high upfront cost but low maintenance cost. Ford is developing a self-diagnostic program for its industrial robots capable of detecting obvious downtime or more subtle maladies via embedded sensors. Cloos International has already implemented a system called Cloos Advanced Robot Diagnostic (CARD) which monitors the entire robot welding system. The following is from Cloos product literature:

> During online operation, the CARD system alerts the operator to relevant problem areas in order to eliminate faults quickly and easily. Help is provided in the form of remedial measures, electrical and mechanical drawings from the service manual, and corrective procedures.
>
> In important areas, the system can anticipate problems by examining the trend of relevant signals comparing them to a historical database. During normal operation, a CAD overview of the complete system is displayed on the PC screen. However, if a problem was to occur in the robot system, relevant screens of helpful information are displayed until the problem is resolved. (Cloos 1991)

As anthrobotic systems become more complex the need for self-diagnostic systems, like CARD or military style remove/replace repair procedures, will increase. Systems with 18 to 36 axes of computer controlled motion are well within our grasp in the near future. As they come closer to human complexity they also come closer to human frailty and diagnostics will be required to "cure" the mechanical patients of the future.

# Man-Amplifiers

### Precursors to Anthrobots

A man-amplifying exoskeleton is an articulated powered frame worn by an operator. They are transitional devices, created in the premicroprocessor era, but were displaced by industrial robots before they were perfected. In the man-amplifier a human provides the control system and the mechanical frame provides the power. Joints on the mechanical frame parallel the operator's joints and permit the operator to move freely, while at the same time increasing the load carrying capability of the operator. This enables the operator to have superhuman strength and lift objects several orders of magnitude greater than normally possible.

Man-amplifiers contradict the robotic goal of taking a man out of the loop; instead, man is made a part of the loop and is in direct contact with the task. This direct coupling of the operator to the machine increases the risk to the operator but also provides greater sensitivity. A better application of this technology is in force reflecting telerobots, where the operator wears a powered exoskeleton that "force reflects" what the robot is sensing. The robot then benefits from the operator's senses and the operator receives the benefit of being removed from the scene of operations.

Man-amplifying exoskeletons were first developed at the General Electric Corporation in the 1960s and were capable of giving a person a 1,500 Lbs lift capacity while exerting only 60 Lbs of force (Shaker 1988). However, limitations in joint dexterity, coupled with hydraulic control technology then in its infancy, forced this technology to be shelved. Although 26 degrees-of-freedom were obtained, the structure lacked three degrees-of-freedom wrists. There were also singularities in the hip, ankle and shoulder joints that led to incompatibility between the human wearer and the apparatus. The exoskeleton design was cumbersome, while high power requirements also contributed to the difficulty. At the present time, renewed work has begun at Los Alamos National Laboratory on an upgraded armored fighting "Powered Exoskeleton Suit" for the Marine Corps, dubbed "Pitman."

## Hardyman

Hardyman (Figures 6.51 and 6.52) was the first attempt to mechanically design a man-amplifying exoskeleton using an hydraulically powered, articulated frame worn by an operator (Cloud 1965). An outgrowth of the successful dual arm telerobot "Handyman," it was a joint Army/Navy proposal begun in November 1965 (General Electric 1966). The final report was filed August 30, 1971. It was not inexpensive: over 25,000 shop hours were required for construction. Figure 6.51 shows Hardyman with an operator, although the system was never fully functional.

Hardyman was designed to be worn like a garment. The master inner exoskeleton controlled the powered slave outer exoskeleton. It power assists the wearer's muscles by amplifying them at a ratio of 25:1 (Gilbert 1968 and 1968a). Crude by the standards of today, hydraulic "tickler" sensors produced error signals from the inner exoskeleton to the outer exoskeleton. The operator lifts 250 Lbs, but it feels like 10 Lbs. A capacity of over 5000 Lbs was hoped for. Hardyman weighed in at 1500 Lbs (3300 Kg) and had 30 degrees-of-freedom. The union of man and machine was seen as a logical stepping stone to full autonomous robotics in the days before compact computing power. Low dexterity, dedicated joint types were chosen, probably for lack of more sophisticated designs (Figures 6.53 and 6.54). The biggest problem was the unsuccessful hydromechanical servo system employed in the legs (Mosher 1968a, 1968b, 1970). Unlike the arms, the legs need constant coordination to achieve balance. Unsupported walking was not achieved. Analyses showed that the legs must operate in three different control modes with 12 joints interacting during walking cycles. The interaction of several hydraulic servo joints in series was beyond the analog control technology of the time. Using both legs simultaneously resulted in "violent and uncontrollable motion by the machine." If an operator were subjected to this it would have literally torn him or her to pieces. One 9-jointed arm operating in only one mode was achieved (General Electric 1969). This success was attributed to its functioning as independent subsystems (Fick 1971).

*Figure 6.51 Hardyman Publicity Photo*

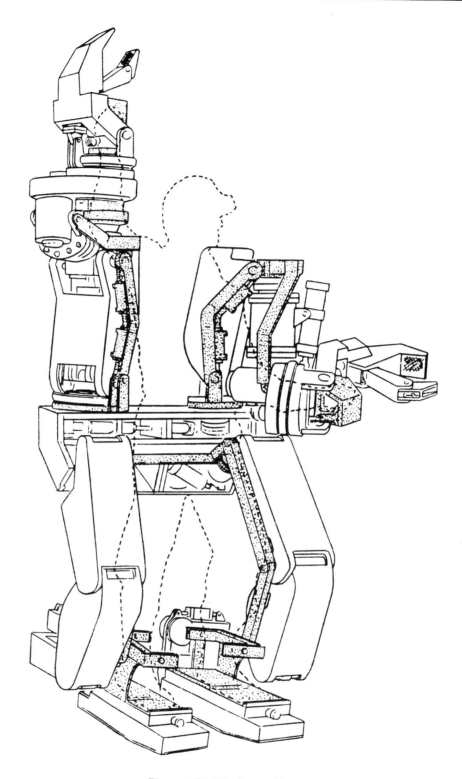

*Figure 6.52 Hardyman Layout*

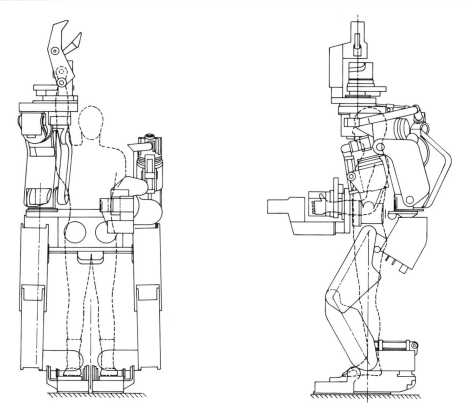

*Figure 6.53  Hardyman Front and Side Views*

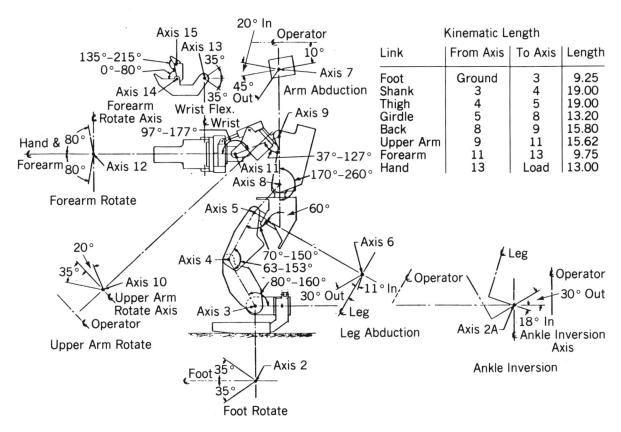

| Kinematic Length | | | |
|---|---|---|---|
| Link | From Axis | To Axis | Length |
| Foot | Ground | 3 | 9.25 |
| Shank | 3 | 4 | 19.00 |
| Thigh | 4 | 5 | 19.00 |
| Girdle | 5 | 8 | 13.20 |
| Back | 8 | 9 | 15.80 |
| Upper Arm | 9 | 11 | 15.62 |
| Forearm | 11 | 13 | 9.75 |
| Hand | 13 | Load | 13.00 |

*Figure 6.54  Range of Motion Study of Hardyman*

### The Man-Amplifier

Figure 6.55 shows a model of the Man-Amplifier proposed by Ross-Hime Designs. It has twenty-six degrees-of-freedom, excluding the hands. A 500 Lbs load capacity is engineered. The modular building block approach applied to the arm and leg actuators simplifies maintenance and reduces parts inventory. Singularity free, pitch-yaw type joints are utilized in legs and arms, an improvement over the simple pivot joints used in the past. The legs have two degrees-of-freedom hips, one degree-of-freedom knees, and two degrees-of-freedom ankle joints. Unique, patented, hydrostatically lubricated ball-and-socket joints powered by ball-screws are used. This design is described in detail in Chapter 5. All joints are powered by electrically driven ball-screw actuators, with the exception of the shoulder and wrist roll axes.

The legs couple to a cross member which supports a three degrees-of-freedom spine (Figure 6.56). The spine gives the human operator's trunk freedom of motion. The spine joint consists of my shoulder joint from Chapter 2 scaled to carry the load of the arms. Consequently it must be more powerful than the shoulder actuators and counterbalanced. Counterbalancing, the negation of the system's own weight, is achieved electronically aided by the self-braking nonbackdriveable nature of the joints. A tubular cross member attached at the top of the spine supports the arms. The arms each have seven degrees-of-freedom; the sum of three degrees-of-freedom shoulder joints, one degree-of-freedom elbows, and three degrees-of-freedom wrists. Both the leg and arm joints are hollow, which permits internal routing of power and position sensing wiring. This is an important feature, as it prevents entanglement and abrasion of the wiring, which could cause increased failures and downtime, not to mention endangering the operator. In "suiting-up," the operator steps backward into the Man-Amplifier and is harnessed to the system via breakaway cuffs located at various lever points on the limbs. A major advantage of the Man-Amplifier is simplified control because the operator is part of the servo loop. As he or she moves, the pressure sensitive cuffs feed a signal to the actuator motors which cause actuator motion.

In contrast to Hardyman, the Man-Amplifier's joint kinematics closely resemble those of the operator. This makes the system inherently user-friendly. The operator does not have to plan for or fight against singularities that might restrict or hurt him. Stability is also improved over past designs, especially in the ankle and hip joints.

Mechanical joint limits are incorporated in the design that prevent harm to the operator and machine. This is particularly important because the numerous degrees-of-freedom and resultant flexibility could force the operator into odd compound joint positions which may conflict with the limits of human joints. Human powered roll rotation might be possible

*Figure 6.55  Man-Amplifier in Emergency Situation*

346

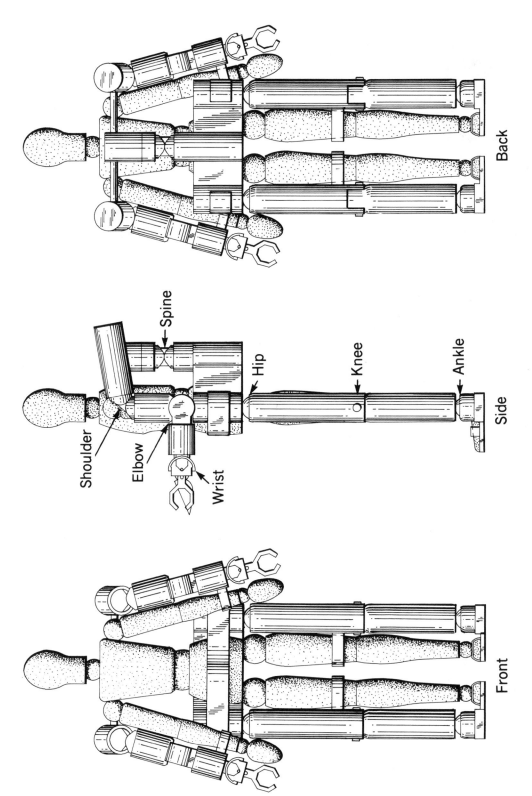

Spine

Shoulder

Elbow

Wrist

Hip

Knee

Ankle

Side

Back

Front

*Figure 6.55 Man-Amplifier from Three Views*

through bearings at various joints such as the ankle. In this case the bearings would take the load, as the minimal rotational loads at such points may well be within the operator's ability to control.

Power for the Man-Amplifier could come from batteries, an umbilical cord, or, as has been proposed with the Los Alamos "Pitman," high efficiency fuel cells. A hydromechanical control system seems logical for simplicity and reliability, but was found to be too cumbersome and inefficient for a mobile device. General Electric has already developed this control system and found it wanting (Mosher 1967, 1971).

Emergency rescue applications include delicate removal of debris from victims entrapped by concrete girders, fallen walls, train wrecks and other man made or natural disasters. The military could find application with clearing shipboard disasters resulting from bombing or missile strikes, aircraft carrier deck fighter plane crashes, etc. Conventional equipment, such as cranes and bulldozers, lack dexterity and sophistication and often endanger the entrapped victims that are being saved. The Man-Amplifier, by virtue of the operator's intimate relationship with the machine, gives great sensitivity to the tasks of disentombing victims from debris. A transference of man-amplifier technology could benefit the physically handicapped population. The partially paralyzed or amputees could regain the ability to walk (Morecki 1984). A full quadriplegic could regain functionality with the appropriate sensor interface. Less exotic uses include manufacturing material management, in which a single operator/man-amplifier team could replace several forklifts.

The advantage of the Man-Amplifier is that it can do the work of many men. It would allow the armed forces to reduce the number of material handling personnel and thereby cut their combat support personnel ratio. Sites where it would be most useful in the near future would be in fortified situations for loading artillery shells and moving supplies or loading cargoes and moving computers when in port. Once its use in support areas is demonstrated, the logical step will be to develop a fast moving fighting machine for rugged terrain. Military applications are wide ranging, and include the augmentation of elite fighting corps (Shaker 1988). Units would be costly, requiring critical applications in which cost is not a factor. As forty percent of the Earth's surface cannot be transversed by wheeled vehicles, the Man-Amplifier's ability to lift and carry heavy equipment such as machine guns, mortars, and ammunition through rugged terrain inaccessible to wheeled vehicles would enable the transport of material on special operations. Construction of emergency shelters and other structures, such as bridges, are other obvious applications for the Man-Amplifier.

## Pitman

Designed as an "enhanced special forces capability for low intensity conflicts and counterterrorists", the Pitman concept is an enhanced version of Hardyman (Figure 6.57). It was conceived at Los Alamos National Laboratory by Jeff Moore of Advanced Weapons Technology Group (Lynch 1991).

The operator is housed in a 500 Lbs fiberglass, polymer/ceramic composite armor called Body Armor, Powered (BAP). Pitman is capable of carrying 300 Lbs of equipment. BAP consists of six layers: (1) Impregnated antibacterial polypropylene; (2) Closed-cell foam laced with heating/cooling tubing; (3) Sealed, impermeable "condom" layer; (4) nitrogen filled polyurethane bubble sensors; (5) energy absorbing closed-cell foam; (6) laminated, composite armor. The exoskeleton frame and joints are lightweight graphite epoxy or Kevlar. Joint structure and kinematics are still unconceptualized at this time. Photoreactive polymer-gel muscles drive the joints. Protected from 0.50 caliber armor piercing rounds, laser, thermal, nuclear, chemical and biological weapons, this 500 Lbs suit appears weightless because the polymer-gel actuators counterbalance the weight (see Chapter 2 for more information on these actuators). The operator views the environment through vibrating mirror virtual displays. Mounted over his head are $C^3I$ Antennae and sensor arrays. An Auxiliary Power Unit (APU)/battery/environmental control pack provides power and cooling/heating for operator comfort.

This design marks a departure from the use of traditional actuation and joint technology. More organic in nature, this monster blends with its operator in as seamless a mechanical interface as possible. Traditional joint technology used in Hardyman and even in the Man-Amplifier may be unsuitable. Actuator control is provided by the soldier's own brain and central nervous system augmented by computers. Lining each segment is a sheet of nitrogen filled polyurethane sacs (similar to bubble wrap). In addition to providing cushioning, this material senses the operator's movements and generates signals that are processed for actuation. In the ultimate system magnetoencephalographic (MEG) sensors would monitor brain activity creating a phantom central nervous system to control the polymer muscles. Hydrogen/oxygen conducting polymer fuel cells supply the power.

Although still years into the future, the Army's interest has been piqued. No doubt in time this technology will become a reality as man pursues bigger and better killing machines (Moore 1986, Pope 1992).

C³I ANTENNAE AND SENSOR ARRAYS

VIBRATING MIRROR VIRTUAL DISPLAYS

APU/BATTERY/ENVIRONMENTAL CONTROL PACK

PHOTOREACTIVE POLYMER-GEL "MUSCLES"

FIBERGLASS POLYMER/CERAMIC COMPOSITE ARMOR

© Aly '91

1. IMPREGNATED, ANTI-BACTERIAL POLYPROPYLENE
2. CLOSED-CELL FOAM LACED WITH HEATING/COOLING TUBING
3. SEALED, IMPERMEABLE "CONDOM" LAYER
4. NITROGEN-FILLED POLYURETHANE BUBBLE SENSORS
5. ENERGY-ABSORBING CLOSED-CELL FOAM
6. LAMINATED, COMPOSITE ARMOR

*Figure 6.57 Pitman Concept Developed at Los Alamos*

## Anthrobots

True anthrobots as defined here are not yet in existence; the present state of AI and control technology requires the use of human beings as the central control and sensing system because nothing as technically sophisticated is available. Man-amplifier technology has been stymied by the lack of appropriately sophisticated control technology. Teleoperators have replaced humans in some hostile environments because the operator can guide the system from a safe area. Although human beings are in the control loop, they are not in contact with the environment or the work

Figure 6.58 shows an experimental master/slave mobile teleoperator. It was developed under the guidance of Raymond Goertz at Argonne National Laboratory during the 1950s, the heyday of American science. A

*Figure 6.58  Argonne Nat'l. Laboratory Mobile Teleoperator*

pair of Model Three electric servo manipulators (five axis mechanical arms) and a stereo, closed-circuit, television system are mounted on an electrically powered wheeled vehicle. The operator (center) is watching the overhead monitor. The slave robot is designed so that one unit can carry out complete repairs on equipment. The machine anticipates the many key features of anthrobots by possessing an entire system of arms as well as mobility.

In the following pages we review the role of corporations, universities, and national laboratories that at various times have brought together the talent and financial resources to create technological milestones. One of the most prominent patrons has been NASA. One of NASA's goals is to develop a robot that performs tasks in a way that is identical to a human. NASA's work began in the 1960s and continues at various centers. NASA's history and involvement in technological areas is so broad that the early efforts have largely been forgotten by the current generation of researchers. Reinventing the wheel is the inevitable outcome of a vast decentralized organization that has lost track of its own early efforts.

## Handyman

Handyman (Figure 6.59) was born in 1958 at General Electric in Schenectady, New York. Built for the Joint AEC-USAF Aircraft Nuclear Propulsion Program (Mosher 1960). Its father, Ralph Mosher, was the same pioneer who would later develop the Walking Truck and Hardyman. A milestone in teleoperator technology. It was: (1) the first servo electro-hydraulic, bilateral master-slave (bilateral servo is one that can be operated at either end with force proportionality), (2) the first articulated exoskeletal master arms that conformed to the operator's arms, and (3) prehensile hands with built-in force reflectors (Mosher 1967). This two armed master-slave manipulator was intended for handling nuclear materials for the Air Force/AEC Nuclear Propulsion Program, with a concrete barrier separating the master from the slave. The electrohydraulic master unit (Figure 6.60) relays the operator's motion through sensors to the electrohydraulic slave. Utilizing the then new servovalve technology it

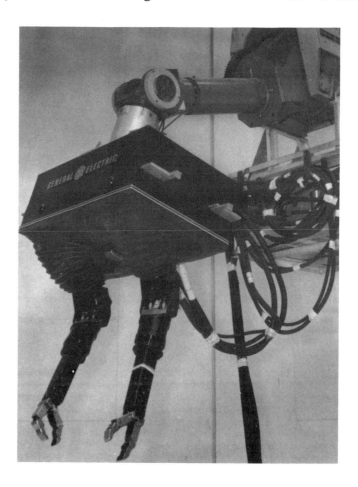

*Figure 6.59  Handyman Built by General Electric*

featured force reflection and counterbalancing of the master arm's weight. Each arm has ten degrees-of-freedom: shoulder, roll-pitch-roll; elbow pitch; wrist roll and pitch; and four finger pitch axis (see Chapter 4 for information on its hands). Fulfilling the roll of the spine it was mounted on a multiaxis crane adding at least two more additional pitch axis (Johnsen 1967, Johnsen 1970). At the debut press conference, Handyman twirled a hoola hoop and wielded a hammer. This was possible by virtue of its force reflecting capability that gave the operator the feel of an object's inertia.

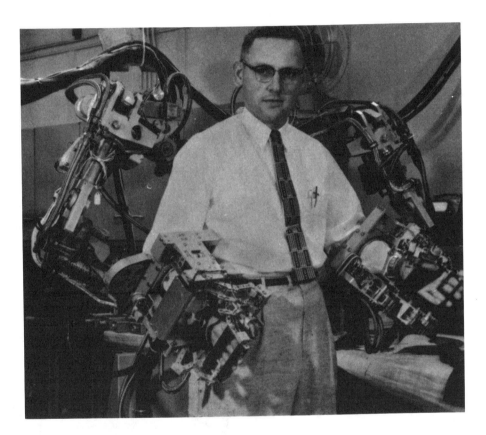

*Figure 6.60  Handyman Master Electrohydraulic Control Unit*

## SAM

The Self-propelled Anthropomorphic Manipulator (SAM) (Figure 6.61) that wears NASA logos was developed under Edwin Johnson's direction in 1969 by the now defunct Space Nuclear Propulsion division of the U.S. Atomic Energy Commission. Johnson is credited with introducing the popular term "teleoperator" in 1966 to describe a servo controlled manipulator that is not directly connected to the operator's twin manipulator. SAM was built at the Nuclear Rocket Development Station at Jackass Flats, Nevada, for inspection and testing of equipment used at radioactive nuclear test sites. Applications envisioned at the time included defusing and disposing of bombs and as search and rescue vehicles in nonnuclear hazardous environments, such as fires or toxic chemicals.

SAM has double arms with a boom capable of traveling vertically and yawing side to side. The torso is mounted on a four-wheeled, dune buggy-like vehicle. A cable provides control signals from an operator wearing an exoskeleton that controls the arms and hands (Figure 6.62). The exoskeleton sends commands to SAM's hands and arms. A closed loop television monitor receives the transmission from SAM's headmounted television camera. This "master/slave" control is one of the basic control methods used in teleoperators, but it was never fully developed (Sullivan 1971).

*Figure 6.61 Self-propelled Anthropomorphic Manipulator*

*Figure 6.62  Exoskeleton Control for SAM*

## Anthropomorphic Master/Slave Manipulator System

The NASA Jet Propulsion Laboratory of Pasadena, California, has developed an anthropomorphic master/slave robot depicted in Figure 6.63. Invented by Hubert C. Vykukal, the U.S. patent was filed on January 24, 1974 (Vykukal 1977). Figure 6.64 shows the master and slave arm sections in operation. As the operator moves the master, the slave duplicates its motions. This design anticipated many of the mechanical features of the Robotics Research Arm discussed in Chapter 2 by utilizing directly driven harmonic drives and sealed tubular construction.

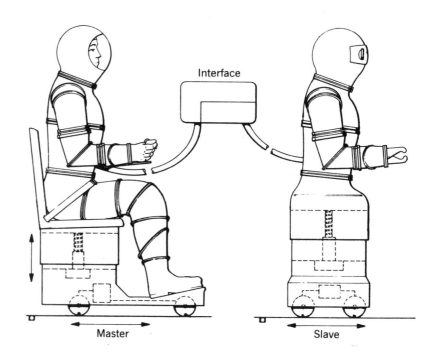

*Figure 6.63  Anthropomorphic Master/Slave*

A joint from the NASA Anthropomorphic Master/Slave Manipulator System Robot resembles an adjustable stovepipe fitting. In each joint all three degrees of freedom intersect at a common center point. Two tubular sections connect at an oblique angle via a ring bearing. Three internal electric motors drive the joint through harmonic speed reducers. For pitch motion, the motor connected to the oblique axis drives the section opposite it. Motors on each end of the wrist provide wrist and tool plate roll. A true up-down pitch motion is obtained by counterrotating the wrist and tool plate roll while rotating the pitch axis.

Harmonic drives were the obvious first choice for powering this type of internal drive joint. Also, NASA's particular configuration, a large number of joints logically added together, makes remote driving impractical and was not pursued. Although simple in implementation, the power in a joint is restricted to the motor size that can be fitted into the joints. Motors in the joint also make it substantially heavier and reduce the load the joint can carry. Response time is also lessened because of the inertial effects of other connected joints.

In operation when the operator dons the master suit and moves the joints or limbs, drive signals are produced in the transducers associated with three bearings that comprise the joint (Figure 6.65). These signals are each applied to the console and are communicated to the slave transducers, which consist of circular bearing assemblies similar to the master suit. These signals, after amplification and phase control connections, cause the torque motors to move each of the respective slave arm portions to an angle that corresponds to the movement of the master arm. A feedback signal indicative of the actual rotation between adjacent tubular portions provides a continuous control on the DC torque motor.

*Figure 6.64 Operation of Master/Slave Sections*

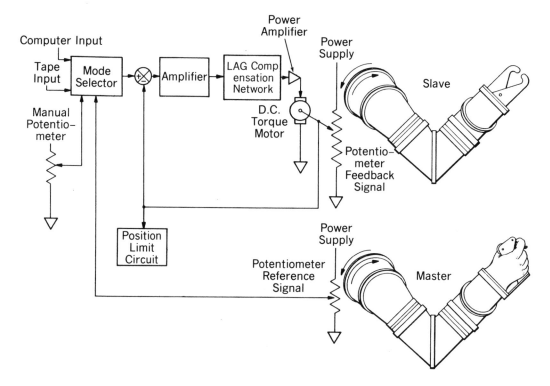

*Figure 6.65  Man-Amplifier Joint Control Block Diagram*

## Green Man

One of the early attempts at simulating the entire upper torso of a human (Figure 6.66) was Green Man, invented by David C. Smith and Frank P. Armogida. It was built during the 1970s by Herbert L. Mummery, M.B. Associates, and Carl Flattau for the Naval Ocean Systems Center (NOSC) in Honolulu, Hawaii, with Navy exploratory development funds. Officially named the Remote Presence Demonstration System, it became known as Green Man when one of its technicians referred to it as the "Little Green Man." Dr. Ross Pepper and David C. Smith at NOSC-Hawaii added stereo cameras, the head mechanism, and the body torso. Its arms, spine, and neck were hydraulically powered, causing occasional "bleeding" by Green Man. Green Man is controlled by a teleoperated exoskeleton harness that mimics the operator's motions much as SAM does. Antagonistic drives using stainless steel cables were experimented with, but abandoned in favor of rotary actuators. Binaural hearing was featured and reportedly worked well. Green Man utilized helmet-mounted display technology with cameras that tilted 45 degrees in any direction. The head tilted down 60 degrees and back 30 degrees in any direction. The basic technology was transferred to the Navy funded Teleoperator/Telepresence Systems (TOPS). Built at the University of Utah, this technology would later form the basis of their entertainment robots and teleoperator arms.

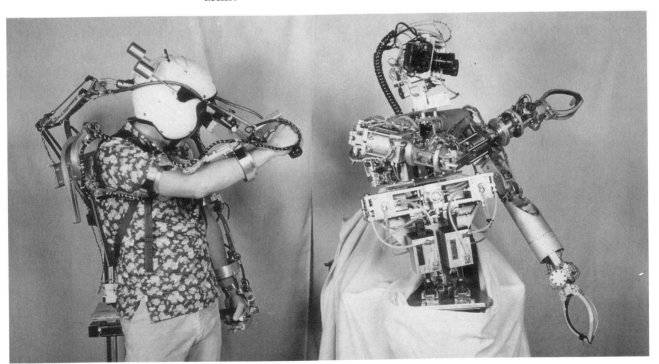

*Figure 6.66 Green Man*

## RETURN OF THE HYDRAULICS?

Recently teleoperated hydraulic robot arms have made a resurgence. Primarily promoted by small companies, their development is due, in part, to their low development cost compared to electric, as well as the readily available and understood hydraulic technology. Two examples are the arms built by the Schilling and Sarcos companies. Both are reminiscent of Handyman, SAM, Green Man, and Manipulating Apparatus from the 1950s (reviewed in Chapter 2).

The Schilling Titan is a five-axis design (Figure 1). Machined out of solid titanium it is very rugged. Hydraulic cylinders provide pitch for the shoulder and elbow. The wrist is a pitch-yaw-roll type created by stacking vane actuators and manifold mounting them together, much like the Manipulator Apparatus. See Chapter 3 for review of this type of wrist. Servo valves are mounted in the forearm. Load capacities in the hundreds of pounds are possible at a low cost and applications have been found in undersea and nuclear fields.

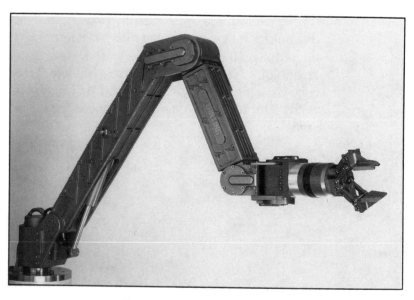

*Figure 1. Schilling Titan II Manipulator*

The more complex ten degree-of-freedom, 3000 PSI hydraulic Sarcos Dextrous Arm uses a roll-pitch-roll shoulder, a one degree-of-freedom elbow, and a single center pitch-yaw-roll wrist (Figure 2). Wrist pitch is limited to 180 degrees, yaw to 90 degrees, and roll is 180 degrees. Miniature servo valves are mounted as close to the actuators as possible. Manifold mounting of the three sizes of vane actuators in the shoulder, elbow and wrists eliminates the need for flexible hoses (Korane 1991).

Load cells located in the arm and end effector can detect weights and loads of less than one ounce providing excellant force reflection sensitivity (Smith 1992). Wiring is external to the arm and is enclosed in conduits and wire guides.

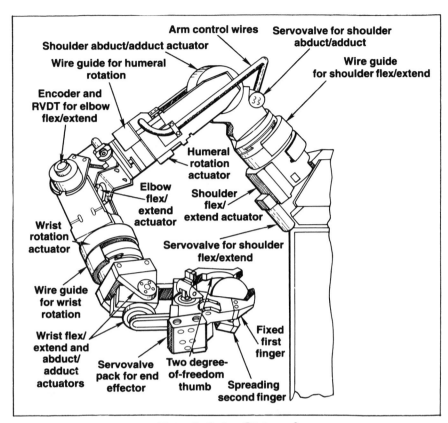

*Figure 2. Sarcos Dextrous Arm*

Designed with single-axis revolute hydraulic components, the Dextrous Arm uses the actuators as joints thus limiting their dexterity and increasing their bulk. The human arm's power is supplied by the muscles, the joint's sole function is to translate the power into dextrous motion. Thus a compact and dextrous design is produced. See Chapter 2 and 3 for comments on similar roll-pitch-roll shoulders and single center pitch-yaw-roll wrists. The hand is reviewed in Chapter 4.

While hydraulic hoses are not used internally, almost all hydraulic arms rely on flexible supply and return lines that connect to a pump and other ancillary equipment. Maintenance of servo valves and other components requires draining the system and inevitably spilling oil. A general decline in hydraulics is evidenced by aircraft control surfaces and automobile brakes that once were solely hydraulic, but now are being designed or replaced with electric actuators. Recently hydraulic oil has been recognized as a "hazardous waste" which makes disposal difficult and costly.

Hydraulic arms with "leak-proof" seals are not new, but have been used by Moog, Inc. in their wrist that employs miniature servo valves. MTS Systems, of Minneapolis, Minnesota, sold a six degree-of-freedom arm, with internally ported hydraulics (Langer 1985). Neither design was a commerical success, not because of poor engineering but because of problems intrinsic with hydraulics in general.

## Shakey

Perhaps the father of all autonomous ground vehicles, Shakey, so named because of its erratic motion, was built at Stanford Research Institute in 1968 (Figure 6.67) under the direction of Charles A. Ross. It was one of the first electrically powered robots to function autonomously, although in a somewhat structured environment. Mechanically primitive, it was able to navigate about a room strewn with geometrical solids. Propulsion was provided by two motorized wheels and sensor bumpers (devices that would become standard on AGV's) detected obstacles, turned off the motors and applied the brakes. A motorized camera and range finder comprised the head. The robot acts as an explorer, by sensing and recognizing objects and boundaries, mapping routes, and navigating its course. Edge detection is used to determine whether an object is a cube, wedge or triangular prism. The "reasoning machine" branch of AI used Shakey for applying logic based, problem solving methods to real world tasks. Shakey simplified the environment into geometrical primitives based on a vision system that was crude by today's standards. One of Shakey's tricks was "pushing the block" off a platform. It used a wedge as a ramp, pushed the wedge against the platform and then drove up the ramp and knocked the block off. No genius, Shakey took an hour to locate the block and ramp in a simple scene (Marsh 1985).

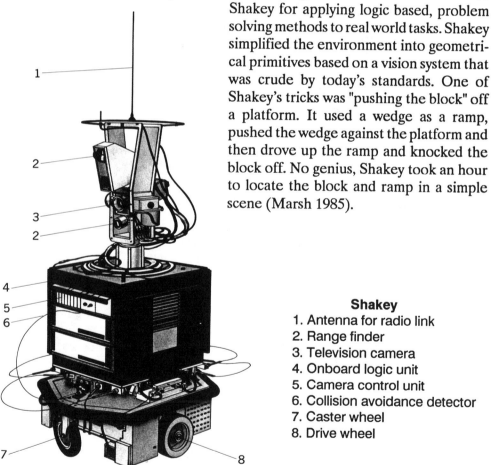

**Shakey**
1. Antenna for radio link
2. Range finder
3. Television camera
4. Onboard logic unit
5. Camera control unit
6. Collision avoidance detector
7. Caster wheel
8. Drive wheel

*Figure 6.67  Shakey*

Shakey's operating program was STRIPS, which constructed plans for Shakey to perform. STRIPS plans were constructed out of primitive robot actions with rules and consequences. Recoverability was provided by incremental replanning, and natural language input was investigated as a means of control. Shakey paved the way for present and future autonomous vehicles. Other vehicles that followed in Shakey's path are JASON, RPI Rover, JPL Rover and the Stanford Cart. One of the largest autonomous vehicle projects is the Martin Marietta Autonomous Land Vehicle (ALV) which can navigate down a paved road at five miles per hour.

The next two pages show the commercial application of Shakey's design twenty years after its inception. Shakey's design is an interesting example of esoteric research laying the foundation for commercial products.

## TRC Helpmate

Developed by robotics pioneer Joe Engelberger, Helpmate (Figure 6.68) is a battery powered autonomous robot designed to pick up hospital food trays. A direct descendant of Shakey, it is guided by passive transponders or bar codes built into the trays. The hospital floor plan is stored in its memory. Navigational sensors include dead reckoning, optical vision, sonar, infrared sensors and structured light to detect obstacles in its travels. A second vision system watches for regular arrays of ceiling lights. Obstacle detection and avoidance algorithms allow it to adjust to its environment.

An auxiliary subsystem handles operator interface functions and a drive subsystem steers the vehicle and provides dead reckoning navigation. All of the sensor subsystems, the drive and auxiliary subsystems, and the 68000 based master processor are linked with a high-speed serial local area network with a master/slave protocol.

Obstacle avoidance is based on the composite local map. Data from the sensor subsystems is rationalized and placed on the map. The map is scanned in front of the robot to see if there are any obstacles. If there are, the robot slows or stops, depending on the distance to the perceived object, and confirms the presence of the obstacle. If the obstacle has moved, as a person will, the robot waits for the path to clear and moves on. If the object is stationary, the robot plans a course around the obstacle, sensing the extent of the obstacle as it moves around it and stays a minimum distance away. Such parameters as search distance ahead, minimum space required for the robot to fit through, speed of motion, speed of turning, and rate of decay of data in the map can be set from a menu system to allow tuning of the robot's behavior (Evens 1989).

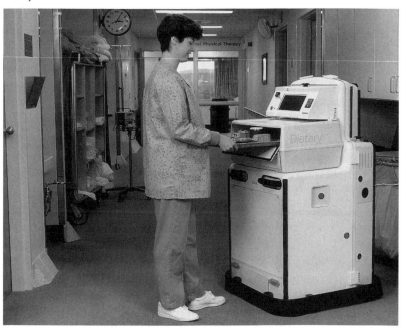

*Figure 6.68  TRC Helpmate*

Through speech synthesis the robot is able to communicate its needs. It sets the tray on the rolling bed tray and delivers it to the patient. On board intelligence and sensors help it to cope with hanging plants, flower arrangements and other, ever changing obstacles. Helpmate's original temperament was very polite, which caused it to miss rides on hospital elevators, so a more pushy behavior pattern was programmed to solve this problem. For more information on Helpmate see the sensors and battery section of this chapter (Feder 1992).

Designs that are the harbingers of anthrobots are examined in the subsequent pages. They provide a glimpse of what is to come. These designs have found ready applications in imitating the human body in controlled test situations, entertaining humans, and by performing tasks in environments hazardous to humans.

**Wasubot**

Invented by Japanese robotics pioneer Ichiro Kato of Waseda University School of Science and Engineering in Tokyo, Wasubot made his debut at Japan's Expo '85 reading sheet music and playing J.S. Bach's *Air on a G String* accompanied by the NHK Symphony Orchestra (Figure 6.69). As with the 18th century automaton the Musician, royalty (in this case the crown prince and princess and emperor of Japan) were on hand to witness this modern day anthrobot performance. Wasubot was also displayed in the Japanese government pavilion along with walking and insect-like robots.

Wasubot is laid out in a generally anthropomorphic shape (Figure 6.70). Its legs function for operating the organ foot pedals and are not capable of walking. In fact the base of its spine is bolted to the bench. Electric motors driving harmonic drives through belts are used through out. Each arm has a roll-pitch-roll shoulder and elbow, and a pitch-yaw-roll wrist. The hands are cable driven and have five digits each. Both hips use roll-pitch-roll actuators and possess knees with pitch-roll ankles.

*Figure 6.69  Wasubot*

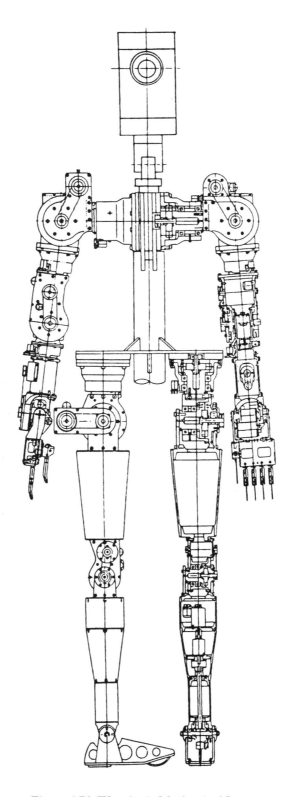

*Figure 6.70 Wasubot's Mechanical Layout*

Spectacular in its dedicated function, Wasubot displays the marriage of machine vision, artificial intelligence, and sophisticated multiaxis control. The kinematics could be improved by additional degrees-of-freedom in the legs, spine and neck. Finally, the cable driven hands as evidenced by the bundles of cables is hardly practical.

Kato describes his goal as follows: "My research is not just in function, but in shape. In thirty years, in the twenty-first century, I think that human form will be essential in robots. In factories, which are for work, robots can be of any shape, but the personal robot, or "My Robot" as I call it, will have to exist in a regular human environment and be able to adjust to humans." Kato is dedicated to replicating animal functions with mechanical systems using his term "biomechantronics." He has also coined the word "synalysis" derived from synthesis and analysis. Kato believes that we must study philosophy and physiology analyzing animal systems to enable us to synthesize mechanical end equivalents (Schodt 1988).

## Robotic Mannequin

Developed by researchers at Pacific Northwest Laboratory (Battele), the robotic mannequin (Figure 6.71) was built for the U.S. Army as an aid in evaluating protective clothing in hostile chemical environments (Yount and Bennett 1987). The robot not only replicates most of the human body's basic motions but also simulates sweating, breathing and skin temperature. The mechanical design has 38 degrees-of-freedom with range of motion and dimensions verified to match the standard physiological values for a 75th percentile male. A library of preprogrammed motion cycles are controlled by the hydraulic power system. The mannequin is tethered by a simple robot arm which allows the mannequin to stand or squat (Figure 6.72). Three layers shield the mechanics and provide various functions (Figure 6.73). A plastic substrata or covering protects the mechanics and provides the anthropomorphic contours. Closed-pore urethane or rubber was used in interference areas such as elbow and knees. Flexible film electrical heaters are overlaid on the substrata to form multiple temperature zones on the body. The skin which overlays the heaters is chlorinated polyethylene cut from sheets and heat sealed at the seams. Tubing terminates in pores at multiple locations to provide perspiration. The polyethylene fabric garment aids in even distribution by allowing the moisture to spread over its surface.

Although a "full range of human motion" is claimed, the joints used in the shoulders do not allow a full range of motion (Figure 6.74). The shoulder uses a roll-pitch-roll joint instead of the human shoulder's pitch-yaw-roll ball-and-socket, which creates a singularity. The singularity occurs when the upper arm is horizontal with the ground plane. What is missing is a true yaw axis of motion, which is provided in the human by the shoulder's ball-and-socket. Shoulder roll is created through a pair of cylinders mounted on a 90 degree out-of-phase double crankshaft. This was necessary to create the 300 degrees of rotation which cannot be obtained from vane actuators. Pitch is produced through a pair of vertically mounted cylinders connecting the upper arm to the shoulder blade area. The second roll actuator, a vane actuator, is mounted near the elbow.

The scapula and clavicle are more faithfully modeled as yaw is provided by a horizontal cylinder mounted in the shoulder blade area. Positioned diagonally, it connects the forward joint assembly to the center column through ball-and-socket tie rods in the upper arm, and the vertically oriented cylinder drives the simulated scapula and clavicle.

The elbow is actuated by a small cylinder. Its range is increased to 210 degrees through a modified bell-crank. The same arrangement is used in the knee. The wrist is a simple pitch-yaw-roll joint driven by two small cylinders with a vane actuator for roll.

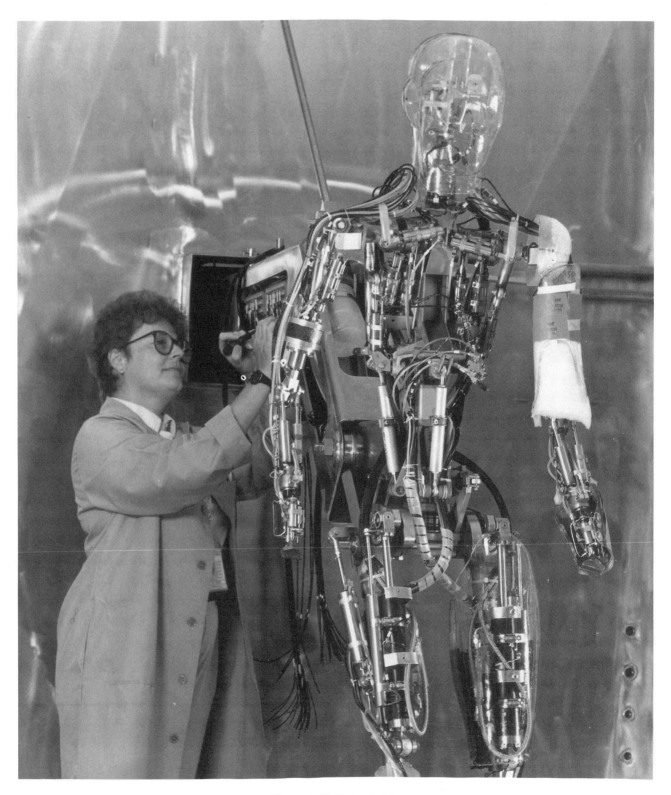

*Figure 6.71  Robotic Mannequin*

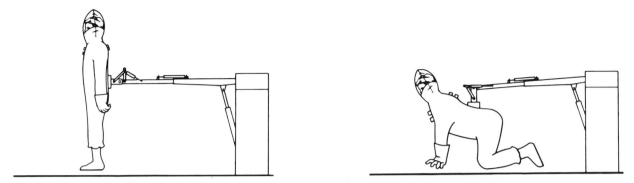

*Figure 6.72  Mannequin Standing and Squatting*

Motion of the spine is simulated by a distended universal joint, in which each axis is driven by a single cylinder. The hips are similar to the spine and utilize a universal joint driven by a pair of cylinders. The pitch range of motion is increased to 210 degrees by a plate type cam. Roll is added through a vane actuator above the knee. Each knee is powered by a pair of cooperating cylinders, and like the elbows and hips uses a modified bell crank mechanism to increase range of motion.

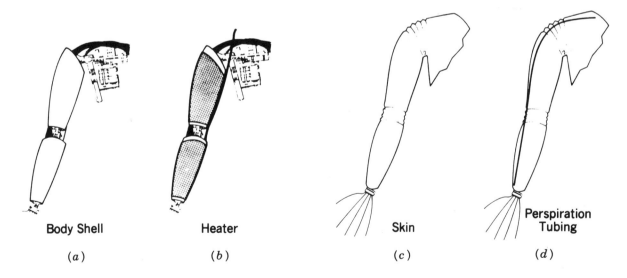

*Figure 6.73  Outer Protective Layers of Robotic Mannequin*

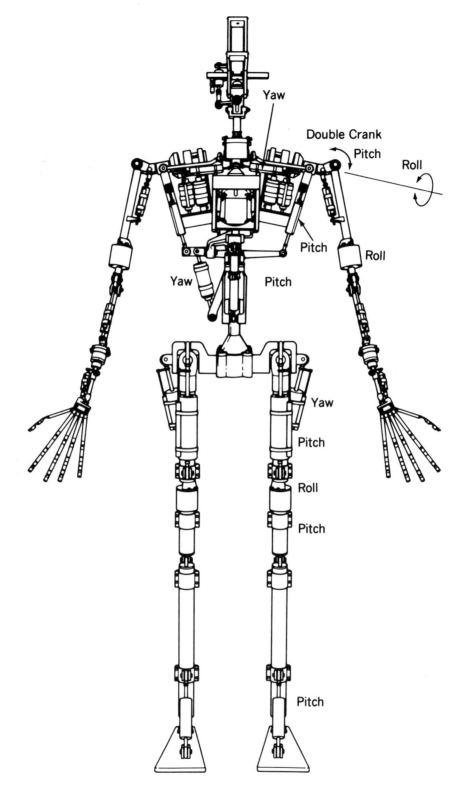

*Figure 6.74 Mechanical Layout of the Robotic Mannequin*

## Sarcos Figures

The Sarcos entertainment figures have up to 56 axes of motion and operate at 3000 psi, utilizing advanced miniature servo valves and high bandwidth actuator technology with programmable compliance. Control strategies for producing graceful motion and programmable figure controllers for recording and playing back motion sequences have been developed. Less mechanically sophisticated than the Robotic Mannequin, the goal is to produce an anthropomorphic appearance (Figures 6.75 and 6.76). The hip has two degrees-of-freedom, front-mounted double acting hydraulic cylinders pitch and yaw are in separate planes. Torso twist is provided by a diagonally mounted cylinder pair. Wrists have three degrees-of-freedom. Small cylinders drive double U-joints for pitch and yaw. The elbow is powered by double acting cylinders encased in the "bicep". The shoulder is roll-pitch-roll, with roll 1 in the torso, pitch in the bicep and roll 2 in the upper arm. In the figures the hydraulic tubing and wiring, similar to arteries and veins, is shown. However, an assemblage of revolute joints does not make for an anthrobotic system. If one analyzes an individual limb such as the arm or a single joint such as the wrist, it becomes apparent that the system does not function precisely like its human counterpart.

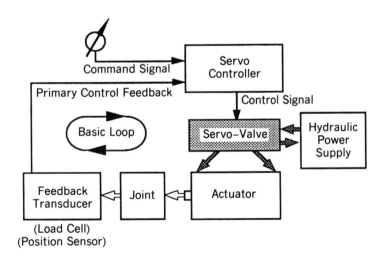

*Figure 6.75 Flow Schematic for Sarcos System*

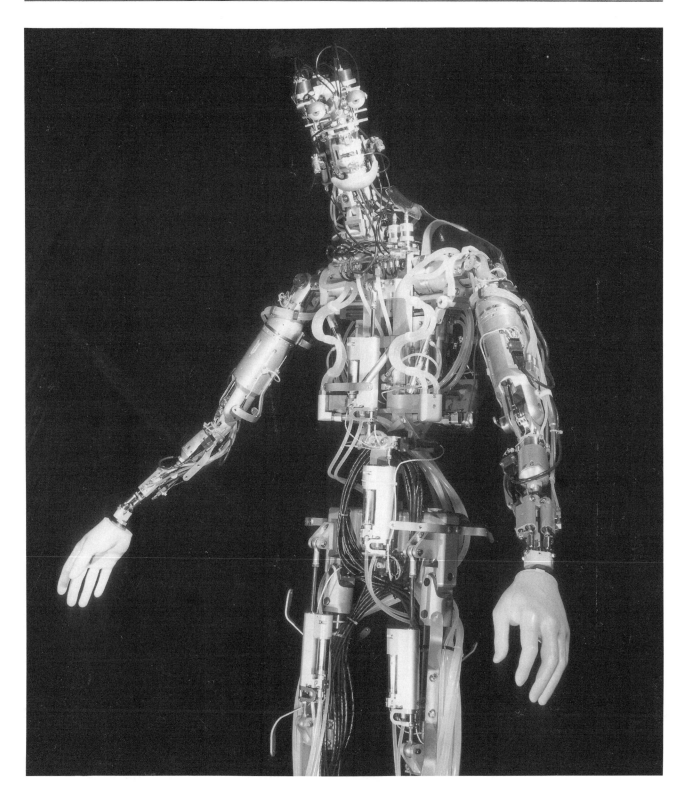

*Figure 6.76  Sarcos Figure*

## EVAR

A direct descendant of Nikola Tesla's robot boat, the Extravehicular Autonomous Robot (EVAR) is being developed at NASA Johnson Space Center (JSC) as a voice supervised, artificially intelligent, free flying robot (Reuter 1988). This anthrobotic Fido will fetch tools or other objects separated from astronauts working on Space Station Freedom (Humphries 1988, McFalls 1988). Figure 6.77 shows a photo montage of EVAR in space. Begun in 1987 as an outgrowth of the "Astrobot" proposal, EVAR is outstanding not so much for its individual components but the synthesis and coordination of these components into a robotic system (Erickson 1991). Although EVAR does not have legs like a human it does have mobility in his environment and so is included in this chapter. A box-like cabinet (Figure 6.78) comprises the body. EVAR is powered by a manned maneuvering unit that produces 7 pounds of thrust. The arms,

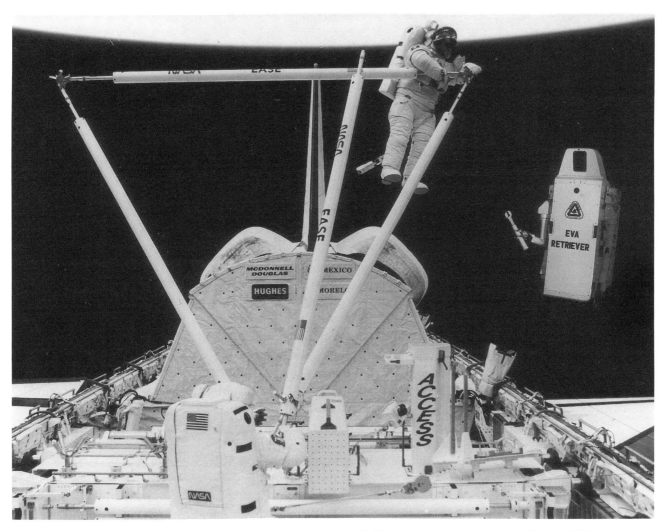

*Figure 6.77 Extravehicular Autonomous Robot (EVAR)*

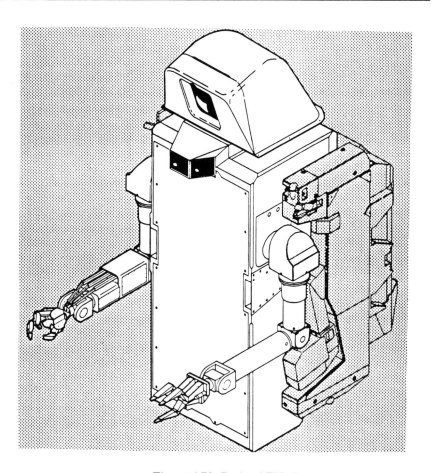

*Figure 6.78  Body of EVAR*

made by Remotec, Inc. Oakridge, Tennessee, have six degrees-of-freedom with singularity ridden pitch roll shoulders, elbows, and pitch-yaw wrists. The Jameson "dextrous" hand described in Chapter 3 is used on one arm with integral optical sensors.

Sensors include accelerometers, gyroscopes, two independent vision systems, and force/proximity sensors on the hands and arms. The accelerometer and gyros measure instantaneous translation and rotational rates of EVAR. The Odetics 3-D laser imaging system (mounted on a turntable) provides primary vision imaging. The second vision system consists of a multiple video camera in the chest.

The processor configuration contains two 15 MHz T4l4 transputers, twelve 20 MHz T800 transputers, a 16 MHz 80386, and six 68020 controllers—a lot of hardware indeed. The transputers are so named because they consist of one microprocessor with four onboard highspeed serial links that communicate with each other (five transputers are dedicated to vision processing). Like transistors these INMOS computers may be used as building blocks for Massive Parallel Processing (MPP). The transputers and 68020s are programmed in C, while the 80386 is programmed in LISP. A "Voice Navigator" by Articulate Systems Corporation is used for voice supervised control with voice synthesis confirmation.

The software architecture incorporates a hierarchical decomposition of the control system that is horizontally partitioned into five major functional subsystems: perception, world model, reasoning, sensing, and acting (Figure 6.79). The design provides for supervised autonomy as the primary mode of operation. Partitioning of the detailed sensing information allows EVAR to focus on the task at hand (foreground), while maintaining the overall picture (background). EVAR dynamically builds its internal environmental knowledge based on continuous sensory perception. There are no preprogrammed environmental models that the environment must conform to. EVAR's capabilities are growing; starting with retrieval of fixed objects, it is rapidly progressing to avoidance of moving obstacles in addition to grasping moving objects.

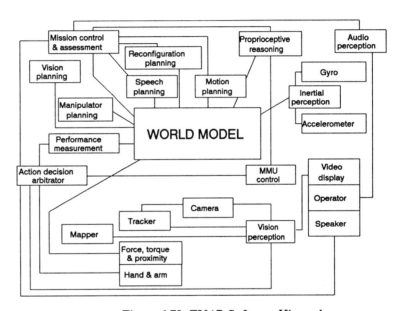

*Figure 6.79 EVAR Software Hierarchy*

## Design Specification

The mechanical translation of human form and function into a workable plan to fulfill the ancient quest for a true man equivalent system is outlined below. The need for dexterity, communication on a human level, and the capability for autonomous action is conceived. Utilizing the technology in each chapter, a basic design departure is made from traditional robotics. In place of identical joint modules for the hips, shoulder, and neck, building blocks that are similar in function but different in shape are used. The shoulder has its own structure as opposed to the wrist, although still kinematically similar. Kinematic primitives are applied to differentiated structures. This is how the human body is constructed.

The mechanical design consists of singularity-free joints. The shoulder, wrist, hips, ankles, spine and neck must be singularity-free to provide maximum dexterity and reduce the computational burden of current singularly ridden joints. Multidigit hands constitute a major subsystem to the singularity-free joint architecture. The anthrobot doesn't have the human body's luxury of large numbers of muscles: in fact, the opposite is sought to minimize the number of actuators needing power and control. This is essential for simplicity, which increases reliability and decreases cost.

Advanced engineered materials are indicated for the anthrobot skeleton of the future. Anthrobots must be optimized to reduce weight, as weight consumes power. Fault tolerant ceramic limbs that may heal if broken is one promising avenue. Actuators, the muscle for the anthrobot of the future, will probably be electric, although not necessarily driven by electric motors. Future development of polymorphic-gel muscles is of interest and may grow to dominance for anthrobot actuator technology.

Mobile power supplies essential to autonomous operation freeing the anthrobot from power cables, cords, or other umbilicals, are needed. Fuel cells seem the most promising power supply currently available. Although expensive, they are clean and relatively compact.

Sensors to provide sight, hearing, touch, body image and proprioceptive data are needed. A gyroscope for balance and stability is essential for highly mobile legged systems which will transverse uneven terrains. Tactile sensors will not only be required for the hands, but could be applied as sheathing to the entire surface of the anthrobot, like the human skin. The polymer film-type sensor described above lends itself to being bonded to three-dimensional surfaces.

Binocular stereo vision using movable and foveated eyes akin to ours will provide depth perception and redundancy. The movable neck will simplify image processing and give the anthrobot another disturbingly human feature. Realtime natural language input and output is required for communicating in the human based world. To support the flood of sensory, actuator, speech recognition and power management input, Massive Parallel Processing (MPP) will be required for understanding and transforming that knowledge into action. Task level programming control initially could come from teleoperation via a force reflecting exoskeleton. The latter probably would be radio frequency linked, as in the ADVANCE system, until MPP computers become compact enough. Onboard computer circuitry may be three-dimensionally interconnected as in the human brain. With human complexity may come human frailties. The vision of the stronger and smarter anthrobot of old may be proven false. High specialization, greater than could be achieved by general purpose

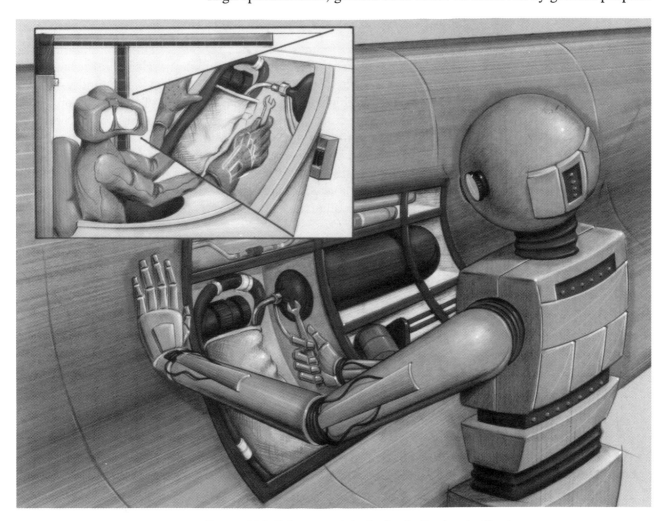

*Figure 6.80  Teleoperated Anthrobot in Service Application*

humans, may be what is ultimately gained. Built-In-Test (BIT) will be necessary to monitor and maintain the multiple electrical, electronic, mechanical systems and subsystems.

Applications for anthrobots include extravehicular activity in outer space such as satellite servicing (Figure 6.80). Space structure assembly is another here-and-now application. A newer application that has been proposed by Cliff Hess of NASA Johnson Space Center is using anthrobots for setting up base camps on other worlds such as Mars and the Moon. This would free the astronauts from the mundane job of interplanetary carpentry. Ross-Hime Designs has been contracted by NASA to build a dual-arm "Robotic Surrogate" to perform these functions.

Astronauts in suits have low dexterity due to suit swelling caused by the vacuum of outerspace and multiple suit layers needed to prevent meteorite penetration. An anthrobot has none of these restrictions and also doesn't need life support other than power to function. It is inherently more dextrous than the astronaut as it is unencumbered by a pressurized suit.

The concept of domestic service robots is gaining popularity. The servant of the future will vacuum and scrub floors, clean bathrooms, wash windows, set and clear tables, cook convenience foods, do laundry, provide security, and possibly maintain appliances. Many of these appliances are already highly structured. Perhaps someday anthrobot compatible appliances, fixtures and even houses will be marketed. Most likely wheeled, rather than legged, it will have many anthrobotic mechanical characteristics including singularity-free shoulders, wrists and perhaps eventually a waist. It will be hoisted from floor to floor like a vacuum cleaner, and will work all day or all night.

# *Epilogue*

Man's desire to depict himself, from cave paintings to the present effort to build a mechanical replica, speaks of his desire to understand his own existence and ease his burden. From ancient figures to the life-like clockwork wonders of the Renaissance and Baroque, this quest continues to the present day. Although the names are changed (anthrobots, automaton, robot, android) they all have in common the goal to create man-like systems. The melding together of human form present in the earliest designs with modern functionality will produce the future synthesis known as anthrobots. Although the motivations differ throughout the ages, the supremacy of man's design never ceases to fascinate and beg emulation.

The evolution of robotics to anthrobotic systems is seen in the early attempts by NASA to harness existing technology for teleoperators, but substituting human control for the lack of artificial intelligence. Designers, whether consciously or unconsciously, are progressing towards anthrobotic systems. The anthropomorphic mannequin shows an anthrobotic system with a solid application for testing clothing. In the TRC Helpmate we have basic artificial intelligence for navigation and maneuverability.

Complete anthrobots may come when work on MPP computers and neural networks matures. Further rationalization is also needed in joint technology to offer a genuine human-like level of dexterity and to create a general mechanical solution for a broad range of applications. Attempts at achieving human levels of dexterity have been frustrated by lack of specialized joint technology. The anthrobot proposed above is the culmination of mechanically designing joints that produce a machine that fulfills the age-old quest for a mechanical counterpart of man, a mirror held up to nature.

The trend is towards mechanical rationalization mentioned at the beginning of Chapter 2; single-purpose joints evolve into multi-purpose joints. It is foreseeable that there will be similar trends in anthrobots, as optimal control and other systems emerge. Progress is being made on higher levels of generality and abstract design principles such as modularity, speed/precision, and load/dexterity. Computers too, have followed a similar evolutionary path, growing from dedicated single-function machines to diverse multi-application general purpose aids.

There are other basic questions that will be encountered in the creation of a man-equivalent mechanical device. Will it be physically more powerful than humans? What about mentally? What are the implications of this, both ethically and practically? Will fail-safe devices be sophisticated enough for their highly sophisticated hosts? How long will they be restricted to specialized hostile environments before they start walking the streets?

Ethical objections abound. When we play God we must also be willing to accept the Creator's burden. Do we really want this? When does an anthrobot stop being a machine and make the transition to being human? ........In Issac Asimov's "The Bicentennial Man" (Asimov 1983) we find an examination of the morality of making something with human capacity for self-reflection. Andrew, the robot-hero fights to exercise a human's rights: the right to wear clothing, the right to earn money, and even the right to die. The growth and journey of Andrew asks us to distinguish the difference between a highly intelligent robot and a man. Will robots become a new species of life? Will additional laws have to be added to Asimov's original three? Does unplugging a robot equal murder? Andrew shows artistic talent, carving wood, and thereby demonstrates a quintessential human trait, creativity. Encouraged by his owner, he becomes wealthy and buys his freedom, much in the fashion of 18th-century indentured servants. While Andrew is constantly modifying himself to increase his resemblance to humans, the factory where he was produced realizes the unpredictable nature of his "generalized pathways" and discontinues his design.

Society begins to see robots as needing a "specialization so fiercely desired by humanity that few robots were any longer independently brained." Robots are controlled through central computers and individual traits become impossible. Humans eventually limit the life of robots to 25 years and human-looking robots are outlawed. In a trial for Andrew's freedom the prosecutor declares: "The word 'freedom' has no meaning when applied to a robot. Only a human being can be free." Freedom may be perceived as a uniquely human passion: only a person can enjoy freedom. However, the judge renders the following ruling: "There is no right to deny freedom to any object with a mind advanced enough to grasp the concept and desire the state." Andrew begins to wear clothes. As with most of Asimov's robot stories, the "Three Laws" and the unique problems they create come into play. When two delinquents order Andrew to take himself apart, the second (the law to obey humans) law supersedes the third law (self-preservation) and exposes the problem with rule-based systems. Fortunately a descendent of Andrew's original owner saves him.

In the end Andrew invents an improved prosthetic device that makes mankind partially robotic and further blurs the difference between man and machine. Ultimately he sees that his own immortality must be sacrificed so that he can achieve what his creators, in their quest for mechanical perfection, sought to escape—being human. Ultimately, the gift that anthrobots give to humans may not be the longed-for escape from the human condition, but rather a reappraisal of what it is we value most about ourselves.

# *Glossary*

**Actuator**  Mechanisms that produce motion.

**Adduction/Abduction**  To bring a limb or other body part towards (Adduction) or away from (abduction) the body.

**Antagonistic relationship**  Pitting of a pair of tendons against each other or other, a means of driving a mechanism that eliminates backlash.

**Accuracy**  The measure of error.

**Android**  Robot in human form.

**Animatronic**  Mid to late 20th century term applied to amusement park robots of human or animal appearance.

**Anthrobot**  Robot that physically and functionally imitates a human.

**Anthropoid**  Human or ape-like morphology having a endoskeleton structure.

**Anthropomorphic**  Having a human-like shape or form.

**Arthropod**  Insect-like exoskeleton structure with generally low dexterity.

**Automaton**  Automatic device with the appearance of a human or other animal. Usually applied to a device from the ancient or medieval time period.

**Back-drive**  Operating a mechanism by driving its output.

**Backlash**  Play or slop in mechanical systems such as bearings and gears. The shortest distance between nondriving tooth surfaces or adjacent teeth in meshing gears.

**Body Image**  A robot's sense of self in time and space. Created through sensor inputs, i.e., gyros, tactile, encoders, etc.

**Circumduction**  A circular pitch/yaw motion of the mechanical wrist's tool plate output shaft.

**Clavicle**  Triangular shaped bone that forms the shoulder blade.

**Closed loop**  A type of servo system in which the output is constantly compared with the input through feedback. The error or difference between the two quantities is used to bring about the desired amount of control.

**Compliance**  Describes flexible systems that "give" or comply to physical constraints.

**Constant velocity joint**  A mechanical universal-type joint that outputs a constant rotational velocity regardless of the joint angulation.

**Cybernetics**  Comparative study of the operations of complex computers and the human nervous system.

**Dexterity**  A combination of range of motion and location of singularities in the work volume.

**Dorsiflexion/Plantarflexion**  Rotation of the foot along the centerline of the leg.

**Endeffector**  The tooling (i.e., grippers, paint guns) attached to a wrist tool plate.

**External rotation**  Rotation of the upper arm away from the body.

**Expert System**  Flexible interactive computer system designed to replicate human problem solving methods.

**Feedback**  Part of the signal from an active circuit or device that is transferred back to the input, such as an error signal, for improving control.

**Flexion/extension** Movement of the wrist by rotating the palm up and down.

**Frames** Objects, events, or situations that are stored so the system can extrapolate consistent data and apply it to similar, yet new, applications.

**Gaits** Manner of walking or running of a creature.

**Gimbal lock** Binding of universal-type joint mechanism by over extending its range of motion or operating in a singularity zone.

**Humerus** Bone connecting the knee to the hip.

**Internal rotation** Rotation of the upper arm towards the body.

**Inversion/eversion** Being turned inward of a part such as a foot, a turning or being turned outward. Yaw motion of the ankle.

**Jacquemart** Medieval figure in clocks that strikes the bell. Usually a knight or solder but may also include animal or mythical figures.

**Lead-through teaching** Method by physically manipulating a robot through a task for programming purposes.

**Log Polar** Mapping function employed in vision; the radial component in the human brain's polar coordinate system is logarithmic, simplifying turning, rotation and zoom into simple data shifts.

**Man-Amplifier** A powered exoskeleton that amplifies the user's strength.

**Magnetostrictive** The phenomenon of certain materials to constrict or shrink in a powerful magnetic field.

**Master/slave control** Control of a robot by a remotely located facsimile of the robot. This may take the form of a joystick to a full exoskeleton.

**Minimax Theorem** A form of data acquisition fundamental to game theory. A method whereby the computer attempts to maximize its score and the opponent tries to minimize the computer's score on its turn.

**Neural networks** A form of computer based on the human brain's neuron structure. Fewer rigid programming techniques are required compared with conventional digital computers.

**Physical Imitation Test**  Method of determining physical human-like quality of robot. If the human is unable to feel the difference in a handshake or other form of physical contact in a subject then it is functionally human.

**Pitch**  Up-and-down motion of the limb.

**Precision**  A combination of repeatability and accuracy.

**Pronation**  To rotate the hand or forearm so that the palm faces down or towards the body.

**Radial/ulnar deviation**  Side-to-side rotation of the palm by the wrist.

**Repeatability**  The measure of error in returning to the same position.

**Robot**  20th century term originally meaning any mechanical man, also any automatic device controlled by computer or remote control.

**Roll**  Rotation of the joint in an axis that is in-line with the arm.

**Self-Diagnostics**  Integral ability to detect and diagnose internal problems. Also known as health and status monitoring, built-in-test.

**SCARA**  Self Compliant Automatic Robot Assembly. Usually configured in a Japanese folding screen design.

**Scapula**  Horizontal bone that connects the shoulder blade to the sternum.

**Semantic Networks**  Use of linking tools such as IF-THEN to form a more efficient method of representing a knowledge base than logical formulas.

**Servo Mechanism**  An automatic control system in which the output is constantly or intermittently compared with the input through feedback so that the error or difference between the two quantities can be used to bring about the desired amount of control.

**Singularity**  The area in a robotic joint's range of motion that has to be avoided because of mechanical or control defects.

**Structured Light**  Projection of light pattern onto a object to aid in computer vision by elimination of extraneous and or conflicting detail.

**Supination/Pronation** Rotation of the hand or forearm upward or away (supination) or downward and towards (pronation) the body.

**Supination** To rotate the hand or forearm so that the palm faces upward or away from the body.

**Teleautomata** 19th century term for a remote controlled device, i.e., a boat or wheeled vehicle.

**Teleoperation** The remote controlling of a device.

**Telerobot** A machine that combines robot attributes with teleoperation control.

**Tool plate** A mounting surface on the arm of the robot wrist for attaching gripper and other endeffectors.

**Turing Test (communication test)** Method of determining the human-like quality of artificial intelligence. If a human is unable to tell the difference in communication with a computer, it is functionally human.

**Triordinate shafts** Three torque tubes concentrically stacked inside one another.

**Vestibular System** Located in the inner ear, it provides humans with a sense of balance.

**Von Neumann Architecture** Fundamental design of all computers: organizes the computer into three main components: control, central processing unit, and memory.

**Walk-through programming** A method of programming a robot by physically leading the joints.

**Yaw** Side-to-side motion of the robot joint.

# References

Akeel, H. 1987. U.S. Patent 4,708,580. "Mechanical Wrist Mechanism." Filed 10 November 1986 and issued 24 November 1987.

Akeel, H. et al., 1989. U.S. Patent 4,807,486. "Three-Axes Wrist Mechanism. Filed 9 November 1987 and issued 28 February 1989.

Albus, J. et al., 1987. "NASA/NBS Standard Reference Model for Telerobot Control System Architecture (NASREM), National Bureau of Standards, Technical Note 1235, June 1987.

Aleksander, I. and Burnett, P. 1987. *Thinking Machines*. New York: Knopf. 208 pages. Illustrated.

Aleksander, I. and Morton, H. 1991. *An Introduction to Neural Computing*. New York: Chapman & Hall. 240 pages. Illustrated.

Anderson, V. 1970. U.S. Patent 3,497,083. "Tensor Arm Manipulator." Filed 10 May 1968 and issued 24 February 1970.

Anon. 1990. "NASA's Planetary Rover Project: Building "Smart" Robots for Space Exploration." *NASA Tech Briefs*. 14:9 (1990). Pages. 10-12.

Anon. 1985. "Going Where Others Have Not Gone Before: The Revolutionary Spine Robot Has Now Entered the Very Competitive Spray Painting Market." *Industrial Robot* 12:1. March 1985. Pages 36-37.

Asada, H. and Takeo, K. 1984. U.S. Patent 4,425,818. "Robotic Manipulator." Filed 30 September 1981 and issued 17 January 1984.

Asada, H. and Toumi, K. 1987. *Direct-Drive Robots: Theory and Practice*. Cambridge Massachusetts: MIT Press. 262 pages. Illustrated.

Asimov, I., et al. 1983. *Machines that Think*. New York: Holt and Company. 627 pages.

Aspray, W. 1990. *John von Neumann and the Origins of Modern Computing*. Cambridge Massachusetts: MIT Press. 376 pages. Illustrated.

Bailey, E. 1989. U.S. Patent 4,827,786. "Preloaded Antibacklash Gear Assembly." Filed 11 February 1987 and issued 9 May 1989.

Ballard, D. 1989. "Frames for Animate Vision." Proceedings of the Eleventh International Joint Conference on Artificial Intelligence. New York: Morgan Kauffman.

Bartholet, S. 1989. U.S. Patent 4,808,064. "Micropositioning Apparatus For Robotics Arm." Filed 5 December 1985 and issued 28 February 1989.

Bartholomew, C. 1979. *Mechanical Toys*. Secaucus New Jersey: Chartwell Books. 143 pages. Illustrated.

Bartlett, D. et al., 1988. U.S. Patent 4,753,128. "Robot with Spring Pivot Balancing Mechanism." Filed 9 March 1987 and issued 28 June 1988.

Belsterling, C. et al. 1985. U.S. Patent 4,536,690. "Tool-Supporting Self-Propelled Robot Platform." Filed 19 October 1982 and issued 20 August 1985.

Bengtsson, B., et al. 1989. U.S. Patent 4,815,911. "Device for Torsion-Proof Connection of an Element in a Robot Arm or the Like." Filed 20 July 1987 and issued 28 March 1989.

Brady, B. and Paul, R., editors. 1984. *The International Journal of Robotics Research* 3:2 (1984). Special Issue on Legged Locomotion.

Brunson, P. et al. 1986. "Next Generation Space Manipulator." Presented at the Conference on Artificial Intelligence for Space Applications. Huntsville, Alabama. November.

Capek, K. 1973. *RUR*, translated by P. Selyer. New York: Washington Square Press.

Carlisle, B., et al. 1987. U.S. Patent 4,702,668. "Direct Drive Robotic System." Filed 24 January 1985 and issued 27 January 1987.

Chaikin, G. and Weiman, C. 1981. U.S. Patent 4,267,573. "Image Processing System." Filed 14 June 1978 and issued 12 May 1981.

Chapuis, A. and Droz, E. 1956. "The Jaquet-Droz Mechanical Puppets." Neuchatel Historical Museum pamphlet. 21 pages. Illustrated.

Chapuis, A. and Droz, E. 1958. *Automata: A Historical and Technological Study*. London, B.T. Batsford Ltd. 408 Pages. Illustrated.

Charlier, M. 1988. "Robots to be Used in Slaughterhouses in Bid to Reduce High Level Injuries." *Wall Street Journal*, 27 December 1988. Page B2.

Cheney, M. 1981. *Tesla: Man out of Time*. Englewood Cliffs, New Jersey: Prentice Hall. 320 pages. Illustrated.

Chun, W. and Brunson, P. 1987. "Actuators for a Space Manipulator." Presented at the 1987 Goddard Conference on Space Applications of Artificial Intelligence and Robotics, 13-14 May 1987, Greenbelt, Maryland.

Chun, W. 1986. "Robotic Lessons Learned and the FTS." Martin Marietta Astronautics. Denver Colorado. 34 pages.

Chun, W., et al. 1989. "Design & Construction of a Quarter Scale Model of the Walking Beam." 20th Pittsburgh Conference on Modeling and Emulation. 5 May 1989. Pittsburgh, Pennsylvania.

Chun, W. 1992. "Planetary Rovers For Space Exploration." *Unmanned Systems* The Magazine of the Association of Unmanned Vehicle Systems. 10:3 (Summer 1992). Pages 26-31.

Clark, K., and Pedretti, C. *1968. The Drawings of Leonardo Da Vinci At Windsor Castle*. Phaidon. New York. Vol. 1. 220 pages.

Clark, R.. 1985. *Works of Man*. New York: Viking Penguin. 352 pages. Illustrated.

Cloos, Inc. 1991. "Cloos Advanced Robot Diagnostic System (CARD)." 1991 promotional literature.

Cloud, W. 1965. "Man Amplifiers: Machines That Let You Carry a Ton." *Popular Science*, November 1965. Pages 70-73 and 204.

Cohen, J. 1967. *Human Robots in Myth and Science*. Cranbury, New Jersey: A.S. Barnes. 156 pages. Illustrated.

Collins, J., et al. 1985. "Death by Robot: Safety Issues in Automated Plants." *Business and Society Review* 54 (Summer 1985). Pages 56-59.

Corwin, M. et al, H. 1975. U.S. Patent 3,920,972. "Method and Apparatus for Programming a Computer Operated Robot Arm." Filed 16 July 1974 and issued 18 November 1975.

Compton, B. 1991. "Neural Networks-the Excitement Continues." OE Reports . Published by SPIE. No. 86. 3 February 1991. Pages 3 and 23.

Critchlow, A. 1985. *Introduction to Robotics*. New York: Macmillan. 491 pages. Illustrated.

Dahlquist, H. 1987. U.S. Patent 4,703,157. "Robot Wrist." Filed 30 June 1986 and issued 27 October 1987.

Dahlquist, H. and Kaufmann, H. 1987. U.S. Patent 4,690,012. "Robot Wrist." Filed 7 May 1986 and issued September 1, 1987.

Department of Energy. 1990. Office of Environmental Restoration and Waste Management, and Office of Technology Development. Environmental Restoration and Waste Management Robotics Technology Development Program. Robotics 5-Year Program. 3 Vols.

Devol, G. 1961. U.S. Patent 2,988,237. "Programmed Article Transfer." Filed December 10, 1954 and issued June 13, 1961.

Devol, G. 1967. U.S. Patent 3,543,910. "Work-Head Automatic Motions Controls." Filed 19 May 1964 and issued 38 February 1967.

Drachmann, A. G. 1948. *Ktesibios, Philon and Heron: A Study in Ancient Pneumatics*. Copenhagen: Munksgaard.

Drachmann, A. G. 1963. *The Mechanical Technology of Greek and Roman Antiquity*. Copenhagen: Munksgaard. 218 pages. Illustrated.

Droz, E. and Chapuis, A. 1958. *Automata: A Historical and Technological Study*. London: B.T. Batsford Ltd. 408 pages. Illustrated.

Dunne, M., et al. U.S. Patent 3,661,051. "Programmed Manipulator Apparatus." Filed 9 May 1972 and issued 9 May 1972.

Duta, O., et al. U.S. Patent 4,878,393. "Dextrous Spherical Robot Wrist." Filed 27 May 1988 and issued 7 November 1989.

Dwolatzky, B., and Thornton, D. 1986. "The GEC Tetrabot – A Serial-Parallel Topology Robot: Control Design Aspects." Chelmsford, UK: GEC Marconi Research Centre

Engelberger, J. 1989. *Robots In Service*. Cambridge, Massachusetts: MIT Press. 348 pages. Illustrated.

Erickson, R., et al. 1991. Technology Test Results From an Intelligent, Free-Flying Robot for Crew and Equipment Retrieval in Space." Proceedings of the Society of Photo-Optical Instrumentation Engineers Symposium on Cooperative Intelligent Robotic Systems in Space II. 10-14 November 1991. Boston, Massachusetts.

Evens, J,. et al. 1989. "Helpmate: A Service Robot for Health Care." *Industrial Robot* 16:2  June 1989.  Pages 87-89.

Feder, B. 1992. "Robotics Comes Back to Reality." *New York Times* 29 April 1992. Page D1.

Feyerabend, F.  1988. "Systematic Optimisation of a Robot Arm Structure." *Industrial Robot* 15:4 December 1988. Pages  219-222.

Feyerabend, F.  1989.  "Faster Robots  From Weight Reduction." *Industrial Robot* 16:4. December 1989. Pages 221-224.

Fick, B, and Makinson, J. 1971. "Final Report on Hardiman I Prototype for Machine Augmentation of Human Strength and Endurance." General Electric Company, Specialty Material Handling Products Operation, Schenectady, New York. 30 August 1971.

Field, J. and Wright, M. 1985. *Early Gearing: Geared Mechanisms in the Ancient and Medieval World*. London: Science Museum South Kensington. 48 pages. Illustrated.

Fisher, L. 1991. "Fuel Cells: Finally Coming of Age or To Good to Be True?" *New York Times*, 30 June 1991.

Furusho, J. and Sano, A. 1990. "Sensor Based Control of a Nine-Link Biped." *The International Journal of Robotics Research* 9:2 (April 1990). Pages 83-98.

Geduld, H. and Gottesmann R. 1978. *Robots Robots Robots*. Boston: New York Graphic Society. 246 pages.

General Electric. 1966. "Exoskeleton Prototype Project: Appendix to Final Report on Phase I." General Electric Company, Mechanical Equipment Branch, Mechanical Technology Laboratory Research and Development Center, Schenectady, New York. 28 October 1966.

General Electric. 1969. "Hardiman I Arm Test: Hardiman I Prototype Project." General Electric Company, Specialty Materials Handling Products Operation, Schenectady, New York. 31 December 1969.

Gilbert, K. 1966. "Exoskeleton Prototype Project: Appendix X to Final Report on Phase I" General Electric Company, Mechanical Equipment Branch, Mechanical Technology Laboratory Research and Development Center, Schenectady, New York. 28 October 1966. 29 pages.

Gilbert. K. and Callan, P. 1968. "Hardiman I Prototype." General Electric Company, Mechanical Equipment Branch, Mechanical Technology Laboratory Research and Development Center, Schenectady NY. June 1968. 38 pages.

Gill, B. 1987. *Many Masks*. New York: Ballantine. 544 Pages. Illustrated.

Gottschalk, M. 1992. "Dextrous Manipulator Targets Hazardous Environments." *Design News*. 10 February 1992. Pages 140-141.

Gunnarsson, S. 1989. U.S. Patent 4,812,132. "Arrangement for Distributing a Cable Assemblage Between Two Mutually Rotatable Component Parts." Filed 13 May 1987 and issued 14 March 1989.

Hart, I. 1925. *The Mechanical Investigations of Leonardo Da Vinci*. Chapman and Hall Ltd. London. 240 pages.

Hart, I. 1961. *The World of Leonardo da Vinci*. London: Macdonald & Co. 374 pages. Illustrated.

Hartley, J. 1984 "Emphasis on Cheaper Robots." *Industrial Robot*, 10:1. March 1983. Pages 49-54.

Hartley, J. 1984 "Semi-Direct Drive to the Fore." *Industrial Robot*. 11:3. September 1984. Pages 158-161.

Hartsfield, J. 1988. "Smart Hands: Flesh is Inspiration for Next Generation of Mechanical Appendages." *Space News Roundup* NASA Johnson Space Center, 27:35, October 1988. Page 3.

Hayes, J. 1983. *The Genius of Arab Civilization*. Cambridge, Massachusetts: MIT Press. 260 pages. Illustrated.

Hess, C. 1990. U.S. Patent 4,980,626. "Method and Apparatus for Positioning Robotic End Effectors." Filed 10 August 1989 and issued 25 December 1990.

Hillier, M. 1988. *Automata & Mechanical Toys*. London: Bloomsbury Books. 200 pages. Illustrated.

Hodges, A. 1983. *Alan Turing: The Enigma*. New York: Simon & Schuster. 587 pages. Illustrated

Hoffer, W. 1989. "Horror in the Skies." *Popular Mechanics* June 1989. Pages 67 -70, 1115 and 117.

Hohn, R. 1975. U.S. Patent 3,909,600. "Method and Apparatus for Controlling an Automation Along a Predetermined Path." Filed 31 May 1973 and issued 30 September 1975.

Hohn, R. 1977. U.S. Patent 4,011,437. "Method and Apparatus for Compensating for Unprogrammed Changes in Relative Position Between a Machine and Workpiece." Filed 12 September 1975 and issued 8 March 1977.

Hole, J. 1990. Letter to author dated 13 February 1990.

Hollerbach, J. and Hunter, I. 1991. "A Comparative Analysis of Actuator Technologies for Robotics." *Robotics Review II*. Edited by O. Khatib, T. Craig and T. Lozano-Perez. Cambridge, Massachusetts: MIT Press. Pages 1-43.

Holt, W., et al. 1988. U.S. Patent 4,780,047. "Advanced Servo Manipulator." Filed 5 April 1985 and issued 25 October 1988.

Humphries, K. 1988. "EVA Retriever: Robot Doesn't Wag Its Tail, But Boy Does it Fetch." *Space News Roundup*. Vol. 27:8, page 3. NASA Johnson Space Center. 8 April 1988.

Hyman, A. 1982. *Charles Babbage, Pioneer of the Computer*. Princeton: Princeton, New Jersey: University Press. 287 pages. Illustrated.

Iwatsuka, N., et al. 1986. U.S. Patent 4,566,843. "Multiarticulated Manipulator." Filed 20 September 1983 and issued 28 January 1986.

Jacobsen, S. et al. 1984a. "The Version I Utah/MIT Dextrous Hand." Proceedings of the Second International Symposium of Robotics Research. Uji, Kyoto, Japan, August 1984. Pages 1-8.

Jacobsen, S., et al. 1984b. "The Utah/MIT Dextrous Hand: Work in Progress." *International Journal of Robotics Research*. 3:4 (Winter 1984). Pages 21-50.

Jacobsen, S., et al. 1987. "Tactile Sensing System Design Issues in Machine Manipulation." Presented at the 1987 IEEE International Conference on Robotics and Automation, Raleigh, North Carolina. 30 March - 3 April 1987.

Jacobsen, S., et al. 1988. "Design of the Utah/MIT Dextrous Hand." Proceedings of the IEEE International Conference on Robotics and Automation, San Francisco, California, 7-10 April 1988.

Jacobsen, S., et al. 1991. "High Performance, High Dexterity, Force Reflective Teleoperator II." Robotics and Remote Systems. Edited by M. Jamshidi and P. Eicker. Alburquerque, New Mexico. 25-27 February 1991. Pages 393-402.

Jagger, C. 1986. *The World's Great Clocks & Watches*. New York: Galley Press. 253 pages. Illustrated.

Jamshidi, M. and Eicker, P. 1989. "Robotics and Remote Systems." Proceedings of the Fourth American Nuclear Society Topical Meeting on Robotics and Remote Systems. Albuquerque NM. January 19, 1991.

Johnsen, E. and Corliss, W. 1967. *Teleoperators and Human Augmentation: An AEC-NASA Technology Survey*. Washington DC: U.S. Government Printing Office. 265 pages. Illustrated.

Johnsen, E. and Magee, C. 1970. *Advancements in Teleoperator Systems: An AEC-NASA Technology Utilization Publication*. Springfield VA: Clearinghouse for Federal Scientific and Technical Information. 236 pages. Illustrated.

Jones, G., et al. 1990. U.S. Patent DES 308,213. "Electric Robot" Filed 6 October 1987 and issued 29 May 1990.

Jovananovic, B. 1990. Letter to author regarding robotic work of N. Tesla, dated 15 October 1990.

Karlen, J. et al. 1989. "A 17 Degree of Freedom Dexterous Manipulator." ASME Annual Meeting, Dynamics and Control Issues for Space Manipulators Session. San Francisco California. 10-15 December 1989.

Karlen, J., et al. 1990. U.S. Patent 4,973,215. "Industrial Robot With Servo." Filed 14 February 1989 and issued 27 November 1990.

Karlen, J. 1992. U.S. Patent 5,155,423. "Industrial Robot With Servo." Filed 18 February 1986 and Issued 13 October 1986.

Karlen, J. 1992. Personal communication, 3 April 1992.

Kato, I. 1973. "Bulletin of Science and Engineering Research Laboratory." No. 62. Weseda University, Tokyo.

Kato, I. 1982. *Mechanical Hands Illustrated*. Tokyo: Survey Japan. 214 pages. Illustrated.

Kau, C., et al. 1989. "Design and Implementation of a Vision Processing System." *IEEE Transactions on Industrial Electronics*. 36:1 (February 1989). Pages 25-33.

Kemp, M. and Roberts, J. 1989. *Leonardo da Vinci*. New Haven, Connecticut. Yale University Press. 1989. 246 pages. Illustrated.

Kenward, C. 1957. UK Patent 781465. " Improvements in or Relating to Positioning, or Manipulating Apparatus." Filed 29 March 1954 and Issued 21 August 1957.

Kimura, M. 1987. U.S. Patent 4,712,969. "Expandable and Contractible Arms." Filed 24 August 1984 and issued 15 December 1987.

Klein, C., et al. 1987. "Vision Processing and Foothold Selection for the ASV Walking Machine." *Proceedings of SPIE - The International Society for Optical Engineering, Mobile Robots II*, Vol. 852, Cambridge, Massachusetts, 5-6 November 1987. Pages 195-201.

Kockan, A. 1991. "Day of the Tripod: Demaurex's Three-Legged Robot has the Speed and Agility to Tackle a Demanding Packaging Application." *Industrial Robot* 18:11(1991). Pages 28-29.

Kohli, D. and Sandor, G. 1989. U.S. Patent 4,806,068. "Rotary Linear Actuator for Use in Robotic Manipulators." Filed 30 September 1986 and issued 21 February 1989.

Korane, K. 1991. "Sending a Robot to do a Man's Job: Servohydraulic Robotic Arm has Humanlike Dexterity." *Machine Design* 63:22 (7 November 1991). Pages 46-49.

Kovan, D. 1989. "Reaching Those Remote Corners of Reactors." *Atom*. No. 388 (February 1989). Pages 8-11.

Kroczynski, P. 1987. U.S. Patent 4,674,949. "Robot With Climbing Feet." Filed 10 January 1983 and issued 23 June 1987.

Kuban, D. and Perkins, G. 1989. U.S. Patent 4,883,400. "Dual Arm Master Controller for a Bilateral Servo-Manipulator." Filed 24 August 1988 and issued 28 November 1989.

Kuban, D. and Williams, D. 1990. "Traction-Drive Force Transmission for Telerobotic Joints." Presented at the ASME 21st Biennial Mechanisms Conference on Cams, Gears, Robot and Mechanism Design. Chicago, Illinois. 16-19 September 1990.

Kunsthistorisches Museum. 1966. *Katalog Der Sammlung fuer Plastik und Kunstgewerbe*, Teil II. Wien: Kunsthistorisches Museum.

Ladendorf, K. 1984. "Developing Dexterity: Victory's robot Has Human-Like Hand Movement." *Austin American-Statesman*. 27 May 1984. Pages D1.

Landsberger, S., et al. 1987. U.S. Patent 4,666,362. "Parallel Link Manipulators." Filed 7 May 1985 and issued 19 May 1987.

Langer, W. 1985. U.S. Patent 4,496,279. "Robot Arm and Wrist Assembly." Filed 9 December 1982 and issued 29 January 1985.

Larson, O. and Davidson, C. 1983. U.S. Patent 4,393,728. "Flexible Arm, Particularly a Robot Arm." Filed 23 February l982 and issued 19 July 1983.

Larson, O. and Davidson, C. 1985. U.S. Patent 4,494,417 "Flexible Arm, Particularly a Robot Arm." Filed 29 September 1992 and issued 22 January 1985.

Lee, K. and Arjunan, S. 1987. "A Three Degree-of-Freedom Micro-Motion in Parallel Actuated Manipulator." *Proceedings of the IEEE International Robotics and Automation Conference.* Scottsdale, Arizona. 14-19 May 1987. Pages 1698-1703.

Linbom, T. 1978. U.S. Patent 4,115,684. "Portable, Programmable Manipulator Apparatus." Filed 17 June 1976 and issued 19 September 1978.

Liston, R. and Mosher, R. 1968. "A Versatile Walking Truck." Prepared for Transportation Engineering Conference. ASME-NYAS. Washington, DC. 28-30 October 1968.

Lloyd, G. 1973. *Greek Science After Aristotle.* New York:Norton & Co. 189 pages. Illustrated.

Loughlin, C. 1991. "Bridgestone Corporation Products Use (ACFAS) System Driven by Rubbertuators." *Industrial Robot* 18:1(1991). Pages. 35-36

Luttgens, K. and Wells, K. 1982. *Kinesiology: Scientific Basis of Human Motion.* New York: CBS College Publishing.

Lynch, T. 1991. "Man Magnifier: The Ultimate Paratrooper." *Design News* 47:8. (22 April 1991) . Pages 84-91.

Marsh, P. 1985. Robots. New York: Crescent Books, 160 pages. Illustrated.

Mason, M. and Salisbury, K. 1985. *Robot Hands and the Mechanics of Manipulation.* Cambridge, Massachusetts: MIT Press, 298 pages. Illustrated

Malone, R. 1978. *The Robot Book.* New York: Harvest HBJ Book. 159 pages. Illustrated.

Markoff, J. 1991. "The Creature That Lives in Pittsburgh." *New York Times*, 21 April 1991. Section 3, pages 1 and 6.

Markoff, J. 1991. "Can Machines Think? Humans Match Wits." *New York Times*, 9 November l991,  Section F, pages 1 and 7.

Markoff,  J. 1993.  "Cocktail-Party Conversation-with a Computer." *New York Times* ,  10 January  1983.  Section F, page 5.

Martin, H. and Kuban, D. 1985. *Teleoperated Robotics in Hostile Environments.* Dearborn, Michigan: Robotics Institute of Society of Manufacturing Engineers. 273 pages. Illustrated.

Martin, H. and Kuban, D.  1989. U.S. Patent 4,828,453. "Modular Multimorphic Kinematic Arm Structure and Pitch and Yaw Joint for Same." Filed 21 April 1987 and issued 9 May 1989.

Maurie, K.  and Otto, M. 1980. *The Clockwork Universe: German Clocks and Automata 1550-1650.* Neale Watson Academic Publications. New York. 321 Pages.

McFalls, D. and Franke, E. 1988. "A Robotic Assistant for Space Station Freedom." *Robotics Today* 2:2 (1988). Pages 1-6.

McGhee, R. 1976. "Robot Locomotion." *In Neural Control of Locomotion*, edited by R.N. Herman, S. Grillner, P.S. Stein and D. G. Stuart. New York: Plenum Press. Pages 237-264.

McGhee, R, et al. 1984. "A Hierarchically Structured System for Computer Control of a Hexapod Walking Machine." *Theory and Practice of Robots and Manipulators. Proceedings of RoManSy 1984: The Fifth CISM-IFToMM Symposium.* Edited by A. Morecki, G. Bianchi and K Kedzior. Kogan Page, London: Hermes Publishing. Pages 375-381.

McGhee, R. and Waldron, K. 1985. "The Adaptive Suspension Vehicle Project: A Case Study in the Development of an Advanced Concept for Land Locomotion." *Unmanned Systems* (Summer 1985). Pages 34-43.

McKinney, W. 1988. "Kinematics of Hooke Universal Joint." NASA Technical Memorandum 100567. June 1988.

Meintel, A. 1990. Personal communication regarding LTM. 11 February 1991.

Meurer, M. 1990. "Selecting Sealed Lead-Acid Batteries." *Powertechnics Magazine* (December 1990). Pages 28-33.

Milenkovic, V. and Hauang, B. 1983. "Kinematics of a Major Robot Linkage." *13th Annual Symposium on Industrial Robots*. Chicago: Society of Manufacturing Engineers. 1983. Pages 31-47.

Milenkovic, V. 1988. U.S. Patent 4,744,264. "Hollow Arm Singular Robot Wrist." Filed 8 July 1985 and issued 17 May 1998.

Milenkovic, V. 1990. U.S. Patent 4,907,937. "Non-Singular Industrial Robot Wrist." Filed 8 July 1985 and issued 13 March 1990.

Miller, D. 1993. "Putting Fuzzy Logic in Focus." *Motion Control*. April 1993. Pages 42-44.

Minium, A. 1918. U.S. Patent 1,265,358. "Combined Driving and Steering Mechanism." Filed 12 May 1916 and issued 7 May 1918.

Miyake, N. 1986. U.S. Patent 4,568,311. "Flexible Wrist Mechanism." Filed 11 January 1983 and issued 4 February 1986.

Moloug, O. 1985. U.S. Patent 4,531,885. "Device for Robot Manipulator." Filed 24 September and issued 30 July 1985.

Mollenhoff, C. 1988. *Atanasoff: Forgotten Father of the Computer*. Ames: Iowa State University Press. 274 pages. Illustrated.

Moore, J. 1986. "Pitman: A Powered Exoskeletal Suit for the Infantryman." Los Alamos National Laboratory publication LA-10761-MS, UC-38. 1986.

Morecki, A., et al. 1984. *Cybernetic Systems of Limb Movements in Man, Animals and Robots*. New York: John Wiley & Sons. 242 pages. Illustrated.

Moses, J. 1992. "Carpal-Tunnel-Syndrome Suits are Consolidated by U.S. Judge. *Wall Street Journal*. 3 June 1992. Page B6.

Mosher, R. 1960. "Force Reflection Electrohydraulic Servo Manipulator." Electro-Technology, December 1960.

Mosher, R. 1967. "Handyman to Hardiman." Society of Automotive Engineers publication MS 670088.

Mosher, R. 1967. "Exploring the Potential of a Quadruped." Society of Automotive Engineers publication MS 690191.

Mosher, R. 1968. U.S. Patent 3,378,028. "Pressure Control Valve." Filed 30 March and issued April 16, 1968.

Mosher, R., et al. 1968. "Exoskeleton Prototype Project: Final Report on Phase I." General Electric Company Report No. S-67-1011. NTIS #505120; General Electric Company, Schenectady, New York. October 1968. 72 pages.

Mosher, R. 1970. U.S. Patent 3,608,743. "Material handling Apparatus." Filed 4 May 1970 and issued 28 September 1970.

Mosher, R. 1971. U.S. Patent 3,557,627. "Apparatus for Compensating Inertial Effects." Filed 26 December 1967 and issued 26 January 1971.

Mosher, R. 1971. "Articulating Mechanism." U.S. Patent 3,580,099. Filed 24 September 1969 and issued 25 May 1971.

Mullen, J. 1972. U.S. Patent 3,694,021. "Mechanical Hand." Filed July 31, 1970 and issued September 26, 1972.

Muybridge, E. 1957. *Animals in Motion.* New York: Dover Publications, Inc. First published in 1899.

Nagahama, Y. 1990. U.S. Patent 4,976,165. "Backlash Removing Mechanism for Industrial Robot." Filed 14 July 1989 and issued 11 December 1990.

Naj, A. 1990. "Product Failure: How The U.S. Robots Lost the Market to Japan in Factory Automation." *Wall Street Journal,* November 6, 1990. Pages 1 and 11.

Neumann, K. 1988. U.S. Patent 4,732,525. "Robot." Filed 21 April 1986 and issued 22 March 1988.

Noguchi, F. 1984. U.S. Patent 4,445,184. "Articulated Robot." Filed 13 July 1981 and issued 24 April 1984.

O'Neill J. 1944. *Prodigal Genius*. New York: Washburn. 326 pages.

Olenick, R. 1985. U.S. Patent 4,551,061. "Flexible, Extendible Robot Arm." Filed 18 April 1983 and issued 5 November 1985.

Ord-Hume, A. 1973. *Clockwork Music*. New York: Crown Publishers, Inc. 334 pages. Illustrated.

Paul, R. and Stevenson, C. 1983. "Kinematics of Robot Wrists." *International Journal of Robotic Research*, (Spring 1983). Pages 31-38.

Panofsky, E. 1940. *The Codex Huygens and Leonardo da Vinci's Art Theory*. The Pierpont Morgan Library. Codex M.A. 1139. London.

Pedretti, C. 1981. *Leonardo Architect*. New York: Rizzoli International Publications. 365 pages. Illustrated.

Pellerin, C. 1992. "The Salisbury Hand." *Industrial Robot*. 18:4. April 1991. Pages 25-26.

Petersen, D. 1984. U.S. Patent 4,489,248. "Linear Actuator." Filed 25 July 1983 and issued 18 December 1984.

Pham, D. and Heginbotham, W. 1986. *Robot Grippers*. New York: Springer-Verlag 443 pages. Illustrated.

Pollard, W. 1942. U.S. Patent 2,286,571. "Position Controlling Apparatus." Filed 22 April 1938 and issued 16 June 1942.

Poole, H. 1989. *Fundamentals of Robotics*. New York. Van Nostrand Reinhold. 436 Pages. Illustrated.

Pope, G. 1992. "Power Suits." *Discover* (December 1992). Pages. 67-71

Pramberger, J. 1990. "NASA's Planetary Rover Project: Building "Smart" Robots for Space Exploration." *NASA Tech Briefs*. 14:9 (1990). Pages. 10-12.

Price, D. 1975. *The Antikythera Mechanism: A Calender Computer From ca. 80 BC*. New York: Science History Publications. 70 pages. Illustrated.

Raibert, M., et al. 1989. "Dynamically Stable Legged Locomotion." MIT Artificial Intelligence Laboratory, Technical Report 1179. LL-6 (1989). 203 pages.

Raibert, M. 1990. "Trotting, Pacing and Bounding by a Quadruped Robot." *Journal of Biomechanics* 23, Suppl. 1 (1990). Pages. 79-98.

Raibert, M. 1990. "Legged Robots." in *Artificial Intelligence at MIT Expanding Frontiers*, Volume 2, edited by Patrick Henry Winston with Sarah Alexandra Shellard, Cambridge, Massachusetts: MIT Press, 1990. Pages 149-179.

Raibert, M. and Hodgins, J. 1991. "Adjusting Step Length for Rough Terrain Locomotion." *IEEE Transactions on Robotics and Automation*, 7:3 (June 1991). Pages 289-298.

Raibert, M. and Hodgins, J. 1992. "Robots, Mobile." *Encyclopedia of Artificial Intelligence*, Second Edition. New York: John Wiley & Sons. Pages 1398-1409.

Reuter, G., et al. 1988. "An Intelligent, Free-Flying Robot." Presented at SPIE Symposium on Advances in Intelligent Robotic Systems, Space Station Automation IV, Cambridge, Massachusetts, 6-11 November 1988.

Rivin, E.I. 1988. *Mechanical Design of Robots*. New York: McGraw-Hill. 325 pages. Illustrated.

Roe, J. 1987. *English and American Tool Builders: The Men who Created Machine Tools*. Bradley, Illinois: Lindsay Publications. 315 pages. Illustrated.

Roselund, H. 1944. U.S. Patent 2,344,108. "Means for Moving Spray Guns or Other Devices Through Predetermined Paths." Filed 17 August 1939 and issued 14 March 1944.

Rosheim, M. 1980. U.S. Patent 4,194,437. "Hydraulic Servo Mechanism." Filed 9 March 1978 and issued 25 March 1980.

Rosheim, M. 1981. U.S. Patent 4,296,681. "Fluid Driven Servomechanism." Filed 14 November 1979 and issued 27 October 1981.

Rosheim, M. 1986. U.S. Patent 4,686,866. "Compact Wrist Actuator." Filed 21 January 1986 and issued 18 August 1986.

Rosheim, M. 1988a. "Wrists." *International Encyclopedia of Robotics Applications and Automation*: Edited by R. Dorf. New York: John Wiley & Sons. Pages 1991-1999.

Rosheim, M. 1988b. U.S. Patent 4,723,460. "Robot Wrist Actuator." Filed 27 January 1987 and issued 9 February 1988.

Rosheim, M. 1988c. U.S. Patent 4,729,253. "Wrist Actuator." Filed 8 August 1986 and issued 8 March 1988.

Rosheim, M. 1988d. "Design of a Omni-Directional Wrist and Shoulder." *Proceedings of Robots 12 and Vision '88 Conference*, Volume 2. Detroit, Michigan. 5-9 June 1988.

Rosheim, M. 1989a. "Man-Amplifying Exoskeleton. Pages 402-411. Wolfe, W. and Chun, W. Chairs/Editors. 1989. "Mobile Robots IV." *Proceedings of SPIE - The International Society for Optical Engineering*. Vol. 1195. Philadelphia, Pennsylvania. 6-7 November 1989.

Rosheim, M. 1989b. *Robot Wrist Actuators*. New York: John Wiley & Sons. 288 pages. Illustrated.

Rosheim, M. 1989c. U.S. Patent 4,804,220. "Wrist Tendon Actuator." Filed 20 February 1987 and issued 14 February 1989.

Rosheim, M. 1989d. U.S. Patent 4,821,594. "Robot Joints." Filed 10 June 1988 and issued 18 April 1989.

Rosheim, M. 1990a. U.S. Patent 4,911,033. "Robotic Manipulator." Filed 3 January 1989 and issued 27 March 1990.

Rosheim, M. 1990b. "Design of an Omni-Directional Arm." *Proceedings of the 1990 IEEE International Conference on Robotics and Automation*. Cincinnati, Ohio, 13-18 May 1990. Pages 2162-2167.

Rosheim, M. 1991. U.S. Patent 5,036,724. "Robot Wrist." Filed 30 November 1987 and issued 6 August 1991.

Rosheim, M., et al. 1993. "Eleven Axis Robot Hand Built With Off-The-Shelf Motion Control Products." *Motion Control*. April 1993. Pages 25-28.

Rosheim, M. 1993a. U.S. Patent 5,239,883. "Modular Robot Wrist." Filed 26 September 1991 and issued 31 August 1993.

Ruoff, C. and Salisbury, J. 1991. U.S. Patent 4,921,293. "Multi-Fingered Robotic Hand." Filed 2 April 1982 and issued 12 December 1984.

Russell, M. 1983. "Odex l: The First Functionoid." *Robotics Age* 5:5 (September and October 1983). Pages 12-18.

Sakaguchi, Y., et al. 1987. U.S. Patent 4,689,538. "Driving Device Having Tactility." Filed 11 September 1986 and issued 25 August 1987.

Saveriano, J. 1981. "An Interview with George Devol. *Robotics Age*. November/December 1981. Pages 22-28.

Schodt, F. 1988. *Inside the Robot Kingdom: Japan, Mechatronics, and the Coming Robotopia*. New York: Kodansha International. 256 pages. Illustrated.

Schrieber, R. 1984. "Volvo Chooses Spine Robot for Spray Operations." *Robotics Today*. (Feb. 1984). Page 28.

Schuman, H. 1983. *Leonardo On the Human Body*. New York: Dover. 506 pages. Illustrated.

Scott, P. 1985. *The Robotics Revolution*. New York: Basil Blackwell, Inc. 345 pages. Illustrated.

Scriptar and Ricci, F. 1979. *Androids: The Jaquet-Droz Automatons*. Milan: Scriptar. 93 pages. Illustrated.

Shaker, S. "Robot Marines: A Few Good Machines." *Sea Power* January 1987. Arlington, VA: Burrelles Press Clipping.

Shaker, S. and Wise, A. 1988. *War Without Men: Robots on the Future Battlefield*. New York: Pergamon-Brassey. 196 pages. Illustrated.

Shelef, G. 1989. U.S. Patent 4,819,496. "Six Degrees of Freedom Micromanipulator." Filed 17 November 1987 and issued April 11, 1989.

Shpigel, V. U.S. Patent 5,101,681. "Interlocking-Body Connective Joints." Filed 13 March 1990 and issued 7 April 1992.

Shum, L. 1983. U.S. Patent 4,407,625. "Multi-Arm Robot." Filed 15 May 1981 and issued 4 October 1983.

Simunovic, S. 1986. U.S. Patent 4,569,627. "Robotic Manipulator." Filed 14 June 1984 and issued 11 February 1986.

Skinner, F. 1975. U.S. Patent 1975. "Multiple Prehension Manipulator." Filed 5 March 1974 and issued 18 February 1975.

Skinner, F. 1975. "Designing a Multiple Prehension Manipulator." *Mechanical Engineering*. September. Pages 30-37.

Skinner, F. 1975. "Designing a Multiple Prehension Hands for Assembly Robots." *Innovation, Inc., Fifth International Symposium on Industrial Robots*. September 22-24. Chicago, Illinois.

Skoog, H., et al. 1987. U.S. Patent DES 292,981. "Robot." Filed 13 November 1984 and issued 1 December 1987.

Skoog, H. 1987. U.S. Patent 4,710,092. "Industrial Robot Having Two Gimbal-Ring Type Arranged Swinging Axes. Filed. 10 May 1985 Issued 1 December 1987.

SME (Society of Manufacturing Engineers). 1983. *Industrial Robots*. 1983. Dearborn, Michigan: Society of Manufacturing Engineers. 416 pages, Illustrated.

Smith, F. , et al. 1992. *Journal of Robotic Systems*. 9(2) New York: John Wiley & Sons. Pages 251-260.

Song, S. and Waldron, K. 1989. *Machines that Walk: The Adaptive Suspension Vehicle*. Cambridge, Massachusetts: MIT Press. 314 pages. Illustrated.

Stackhouse, T. 1983. U.S. Patent DES 268,033. "Robot Arm." Filed 22 December 1980 and issued 22 February 1983.

Stackhouse, T. et al. 1987. U.S. Patent D293116. "Industrial Robot." Filed 5 June 1985 and issued 8 December 1987.

Standh, S. 1989. *The History of the Machine*. New York: Dorset Press, 240 pages. Illustrated.

Stauffer, R. 1983. "Intelledex Targets New Robot for Precision Electronics Assembly." *Robotics Today* (April 1983). Pages 67-68.

Stauffer, R. 1984. "Asea's Pendulum Robot." *Robotics Today*. December 1984. Pages 31-32.

Stauffer, R. 1987a. "Trends in Tooling: Smarter, More Versatile, More Tolerant." *Robotics Today* (August 1987). Pages 19-22.

Stauffer, R. 1987b. "Researching Tomorrow's Robots." *Robotics Today* (October 1987). Pages 27-35.

Steffen, C. and Su, R. 1993. "Inverse Kinematic Solution for a Class of Seven Degree-of-Freedom Redundant Manipulators With Double Universal Wrist and Shoulder." Submitted to IEEE Transaction on Robotics and Automation in 1993.

Steffen, C. 1994. "Configuration Space Equivalence and Inverse Kinematic Solutions for Manipulators." PhD Dissertation, Electrical Engineering Department, University of Colorado at Boulder. Boulder Colorado.

Stoughton, R., et al. U.S. Patent 4,762,016. "Robotic Manipulator Having Three Degrees-of-Freedom." Filed 27 March and issued 9 August 1988.

Sullivan, G. 1971. *Rise of the Robots*. New York: Dodd, Mead & Co. 114 pages. Illustrated.

Suica, D., et al. 1990. U.S. Patent DES 307,282. "Industrial Robot." Filed 11 February 1987 and issued 17 April 1990.

Sutherland, I. 1990. U.S. Patent 4,900,218. "Robot Arm Structure." Filed 18 June 1985 and issued 13 February 1990.

Sutherland, I. 1983. "A Walking Robot." Pittsburgh: Marcian Chronicles, Inc. 131 Pages.

Svensson, R., et al. 1990. U.S. Patent 4,904,150. "Balancing Unit for Pivotable Mechanical Elements, such as Doors, Robot Arms, etc." Field 12 May 1988 and issued 27 February 1990.

Takagi, et al. 1986. U.S. Patent 4615,260. "Pneumatic Actuator For Manipulator." Filed 25 April 1984 and issued 7 October 1986.

Tesar, D. 1989. "A Generalized Modular Architecture for Robot Structures." *Manufacturing Review*. 2:2 (June 1989). Pages 91-118.

Tesla, N. 1983. *My Inventions: The Autobiography of Nikola Tesla.* Williston, VT: Hart Brothers. 111 pages. Illustrated.

Tesler, L. 1991. "Networked Computing in the 1990s." *Scientific American* 265:3 (September 1991). Pages 86-93.

Thornton, G. 1986. "The GEC Tetrabot – A New Serial-Parallel Assembly Robot." GEC Marconi Research Centre. Chelmsford, U.K. See also "Development of the 'Tetrabot' Robotic Manipulator." Engineering Research Centre-Whetstone. Techbrief. W/924/T/005.

Todd, D. 1985. *Walking Machines: An Introduction to Legged Robots.* New York: Chapman and Hall 190 pages. Illustrated.

Torrey, H. 1892. *The Philosophy of Descartes: In Extracts from His Writings.* New York: Henry Holt and Company. 351 pages.

Trechsel, H. 1992. U.S. Patent 5,117,700. "Miniature Linear Actuator." Filed 25 February 1991 and issued 2 January 1992.

Trevelyan, J.P. 1986. "A Wrist Without Singular Positions." *Robotics Research.* 4:4. (Winter 1986). Pages 71-85.

Tsuge, K. 1991. U.S. Patent 4,990,050. "Wrist Mechanism." Filed 10 April 1989 and issued 5 February 1991.

Ubhayakar, S. and Baker, R. 1990a. U.S. Patent 4,964,062. "Robotic Arm Systems." Filed 16 February 1988 and issued 16 October 16 1990.

Ubhayakar, S. and Baker, R. 1990b. U.S. Patent 4,954,952. "Robotic Arm Systems." Filed 2 October 1989 and issued 4 September 1990.

Usher, A. 1982. *A History of Mechanical Inventions.* New York: Dover. 450 pages. Illustrated.

Van der Spiegel, J. et al. 1989. "A Foveated Retina-like Sensor Using CCD Technology." *Analog VLSI Implementation of Neural Systems,* edited by C. Mead and M. Ismail. Boston: Kluwer Acad. Pages 189-210.

Venkataraman, S. and Iberall, T. 1990. *Dextrous Robot Hands.* New York: Springer-Verlag. 345 pages. Illustrated

Vertut, J. and Coiffet, P. 1986. "Robot Technology." *Teleoperations and Robotics: Evolution and Development*, Vol 3A. Englewood Cliffs, New Jersey: Prentice Hall. 332 pages. Illustrated.

Vickers, M. 1988. U.S. Patent 4,790718. "Manipulators." Filed 27 March 1986 and issued 13 December 1988.

Vukobratiovic, M. 1975. "Legged Locomotion Robots and Anthropomorphic Mechanisms." Michailo Pupin Institute, Belgrade.

Vykukal, H., et al. 1977. U.S. Patent 4,046,262. "Anthropomorphic Master/Slave Manipulator System." Filed 24 January 1974 and issued 6 September 1977.

Wada, M and Nishihara, K. 1987. U.S. Patent 4,685,349. "Flexibly Foldable Arm." Filed 20 December 1985 and issued 11 August 1987.

Waldron, K., et al. 1984. "Configuration Design of the Adaptive Suspension Vehicle." *The International Journal of Robotics Research*, 3:2 (Summer 1984). Pages 37-47.

Watzman, A. 1990. "Autonomous Mobile Robot for Planetary Exploration is Developed for NASA at Carnegie Mellon University." Press release. 17 May 1990.

Weber, R. 1991. Letter to author regarding storage batteries, dated 5 February 1991.

Weiman, C. 1990. "Log-Polar Vision for Mobile Robot Navigation." Electronic Imaging Conference, November 1990, Boston, Pages 382-385.

Weiner, E. 1989. "Repairs to Older Planes Ordered; Move will Affect 2,200 Airliners" *New York Times*, 19 May 1989. Page 1.

Wells, F.E. 1972. U.S. Patent 3,631,737. "Remote Control Manipulator for Zero-Gravity Environment." Filed 18 September 1970 and issued 4 January 1972.

Wendler, G. 1918. U.S. Patent 1,281,448. "Shaft Coupling." Filed 18 April 1916 and issued 15 October 15 1918.

Weyer, P. 1987. U.S. Patent 4,683,767. "Rotary Actuator with Backlash Elimination." Filed 17 January 1985 and issued 4 August 1987.

Wiener, N. 1965. *Cybernetics: or Control and Communication in the Animal and the Machine.* Cambridge, Massachusetts: MIT Press. 212 pages. Illustrated.

Wilcock, P. 1986. U.S. Patent 4,621,965. "Manipulators." Filed 6 February 1984 and issued 11 November 1986.

Williams, R. 1990. "Forward and Inverse Kinematics of Double Universal Joint Robot Wrist." Presented at the Space Operations, Applications and Research (SOAR) Symposium. Albuquerque NM. 26-28 June 1990.

Williams, R. 1992. "The Double Universal Joint on a Manipulator: Solution of Inverse Position Kinematics and Singularity Analysis." NASA Technical Memorandum 104214. 1992.

Wittwer, F. and Suica, D. 1988. U.S. Patent 4,756,204 "Counterbalance Assembly for Rotatable Robotic Arm and the Like." Filed 11 February 1987 and issued 12 July 1988.

Wood, D., et al. 1988. U.S. Patent 4,784,010 "Electric Robotic Work Unit." Filed 27 April 1987 and issued 15 November 1988.

Woodruff, D. 1990. "Will the 21st Century Be Battery-Operated?" *Business Week.* 24 December 1990. Pages 40-41.

Wright, A. 1985. U.S. Patent 4,555,217. "Robot Arm With Split Wrist Motion." Filed 11 January 1985 and issued 26 November 1985.

Wright, D. and Chun, W. 1989. "Advanced Robot for Planetary Exploration." SPIE Vol. 1007: Mobile Robots III. Cambridge, Massachusetts. 10-11 November 1988. Pages 120-127.

Yasuoka, H. 1986. "Backlash Correcting Device for Swival Arm of Robot." U.S. Patent 4,630,496. Filed 30 January 1985 and issued 23 December 1986.

Yoshiyuki, N. et al. 1984. "Hitachi's Robot Hand." *Robotics Age* 6: 7 (July 1984). Pages 18-20.

Youmans, A. 1958. U.S. Patent 2,861,699. "Method and Apparatus for Performing Operations at a Remote Point." Filed 16 October 1950 and issued 25 November 1958.

Young, F. and Binder, W. 1989. U.S. Patent 4,804,170. "Counterbalance Assembly for Rotatable Robotic Arm and The Like." Filed 11 February 1987 and issued 14 February 1989.

Yount, J. and Bennett, D. 1987. "Design, Construction and Testing of an Anthropomorphic Robot." *17th International Symposium on Industrial Robots - Robots ll Conference Proceedings, Chicago, IL, April 1987.* Dearborn, Michigan: SME. Pages 26 through 30, and 1-77 through 1-86.

Zimmer, E. 1987a. U.S. Patent 4,657,472. "Manipulator Head Assembly." Filed 7 June 1985 and issued 14 April 1987.

Zimmer, E. 1987b. U.S. Patent 4,662,8l5. "Manipulator Head Assembly." Filed 26 July 1984 and issued 5 May 1987.

Zimmer, E. 1987c. U.S. Patent 4,690,012. "Robot Wrist." Filed 7 May 1986 and issued 1 September 1987.

Zimmer, E. 1988. U.S. Patent 4,736,645. "Gear Unit For a Manipulator." Filed 18 July 1986 and issued 12 April 1988.

# *Index*

## A

## B

## C

# *Figure Credits*

I wish to thank the following publishers and companies that have granted permission for reproduction of their illustrations. T=Top, B=Bottom, L=Left, and R=Right.

The History of the Machine. Strandh, S. Copyright 1979 AB Norbok pg. 4. Automata: A Historical and Technological Study. Copyright 1958 Editions du Griffon, Neuchatel. pg. 5L. A History of Mechanical Inventions. Usher A. Copyright 1988 Dover Publications, Inc. pg. 5R. The Antikythera Mechanism: A Calendar Computer From ca. 80 BC. Copyright 1975 Science History Publications pg. 8. Freer Gallery of Art pg. 10. Museum of Fine Arts, Boston. 11 pg. IBM pg 18. Kunsthistorisches Museum pg. 22. Androids. Edited by Ricci F. Copyright 1979 Scriptar SA. pg. 24. Stanford University California Museum of Art pg. 28. Tesla Museum pg. 34. Hand or Simple Turning Principles & Practice. Holtzapffel, J. Copyright 1976 Dover. pg. 38T. Hardinge Brothers, Inc. pg. 38B. Kinesiology: Scientific Basis of Human Motion by Luttgens, K. and Wells, K. Copyright 1982 Wm. C. Brown Communications, Inc. Dubuque, Iowa. pg. 43. Society of Manufacturing Engineers pg. 52. NASA pg. 53. Kinesiology: Scientific Basis of Human Motion by Luttgens, K. and Wells, K. Copyright 1982. Wm. C. Brown Communications, Inc. Dubuque, Iowa. pg. 54. Van Nostrand Reinhold pg. 57. Spine Robotics pg. 58T. Industrial Robot Vol. 18:1 pg. 67. Planet Corporation pg. 78. Robots. Marsh, P. Copyright 1985 Salamander Books Ltd. Rise of the Robots. Sullivan G. Copyright 1971 Dodd, Mead pg. 85. NASA pg. 87. Staubli Unimation pg. 89. Mechanical Design of Robots by Rivin, E. Copyright 1988 McGraw-Hill, Inc. pg. 91. ABB pg. 93. ABB pg. 94. ABB pg. 95. ABB pg. 96T. ABB pg. 97. Fanuc 100. Mechanical Design of Robots by Rivin, E. Copyright 1988 McGraw-Hill, Inc. MIT pg. 103T. University of Santa Barbara. pg. 103B. MIT pg. 105. Machine Design pg. 107. Kinesiology: Scientific Basis of Human Motion by Luttgens, K. and Wells K. Copyright 1982 Wm. C.